Soundscape and the Built Environment

Soundscape and the Built Environment

Edited by
Jian Kang
Brigitte Schulte-Fortkamp

CRC Press
Taylor & Francis Group
Boca Raton London New York

CRC Press is an imprint of the
Taylor & Francis Group, an **informa** business

A SPON PRESS BOOK

CRC Press
Taylor & Francis Group
6000 Broken Sound Parkway NW, Suite 300
Boca Raton, FL 33487-2742

First issued in paperback 2017

© 2016 by Taylor & Francis Group, LLC
CRC Press is an imprint of Taylor & Francis Group, an Informa business

No claim to original U.S. Government works

ISBN-13: 978-1-4822-2631-7 (hbk)
ISBN-13: 978-1-138-89308-5 (pbk)

Visit the Taylor & Francis Web site at
http://www.taylorandfrancis.com

and the CRC Press Web site at
http://www.crcpress.com

Contents

Preface

Recent work in soundscape disciplines provides a new multidisciplinary approach and clearly represents a paradigm shift in environmental noise and annoyance research. It focuses on perception and involves human and social sciences (e.g., psychology, sociology, architecture, anthropology, and medicine) to account for the diversity of soundscapes across countries and cultures. Moreover, it considers environmental noise a resource rather than a waste. Soundscape research is related not only to environmental noise management but also to fields such as wilderness and recreation management, urban and housing design, and landscape planning and management.

This is the first book to systematically discuss soundscape in the built environment. It is based on the 4-year European Cooperation in Science and Technology (COST) Action Soundscape of European Cities and Landscapes, where COST is the European framework supporting transnational cooperation among researchers, engineers, and scholars across Europe. The purpose of the action was to provide the underpinning science and practical guidance in soundscape. The action created a vibrant and productive international network of 52 participants from 23 COST countries and 10 participants outside Europe, and a series of workshops, think tanks, conferences, and training schools were organized (http://soundscape-cost.org).

This book is aimed at professionals, including urban planners, landscape architects, architects, acoustic consultants, environmental consultants, and transport engineers, as well as academics and researchers and students in the fields of architecture, landscape, acoustics, building science and technology, environmental science, urban planning, transport engineering, and environment and behaviors. Other related sectors include local authorities and developers.

This book is reasonably self-contained, with some prior knowledge of elementary acoustics assumed. Thus, it is introduced at an introductory level. However, because a considerable amount of recent research results is included in the book, acoustic researchers and practitioners at an advanced level may also find the book useful.

The book provides the basics of soundscape and its meaning with regard to people's health, tools for implementing the soundscape approach, measurement techniques, mapping, soundscape practice, and examples.

Chapter 1 discusses the acoustic environment and soundscapes in general, including sound sources in the acoustic environment, with a classification scheme for categorizing sound sources in any acoustic environment that can be used to standardize sound source reporting across different studies; scope and matters pertinent to the eventual definition of the soundscape; the soundscape–landscape analogy; soundscape in different languages; the role of context in soundscape assessment; and a comparison between soundscape and environmental noise. An emphasis is placed on soundscape as part of the living environment that can partly be measured and described by physical quantities, and less on soundscape as an artistic expression (composition or installation) or soundscape as a mood or an idea that only exists in the mind of a person.

Chapter 2 discusses the basis of scientific evidence from psychophysics, psychology, cognition, and so forth—how perception, understanding, and judgement (appraisal) emerge when a person is embedded in a given sonic environment. The important roles of auditory scene analysis, listening modes, and attention focusing, as well as the attribution of meaning to the individual sounds, are explained. Various context elements, such as multisensory experience, expectation, discourse resonance, and cultural variations and the way they influence soundscape, are explored. Interaction schemes and models are used to give the reader a deep understanding of soundscape. In particular, the chapter looks at listening modes, auditory scene analysis, attention processes at various levels (multisensory), the influence of expectation on listening, how sounds get meaning (multisensory), what influences appraisal, and how discourse in society influences the process.

Chapter 3 explores the notion that access to high-quality acoustic environments plays an intrinsic role in human well-being, quality of life, and the environmental health of urban and rural populations—through physical, psychological, and physiological restorative mechanisms. This chapter reviews the limited literature available on soundscapes and restoration, and possible models for their interconnection. The review intersects the wider literature of environmental noise and health and publications on well-being and natural green environments. The economic perspective is also included: How much are people willing to pay for scarce goods such as tranquillity and quiet, and what are the benefits? Suggestions are made about where further work is needed to shed more light on any potential relationship between soundscapes and health.

Chapter 4 discusses the meaning of soundscape research in its relevance for social and acoustic harmony in living environments. Soundscape research does not conceive sound alone, but reconceives the condition and purpose of its production, perception, and evaluation, which account for a human-centred point of view. The chapter discusses human perception as the guidance of soundscape, the soundscape approach, approaching people's minds, general soundscape analysis methods, the perception of sound quality, the perception of acoustical environments, the concept of triangulation, and modelling perception with regard to societal needs. Tools for exploration have been evaluated based on the new understanding of expertise to enhance the quality of life.

Chapter 5 analyzes large-scale data on perceived soundscapes from several Dutch and Austrian surveys, because that would be valuable to know how people perceive their sound environment in areas with varying levels of road, air, or rail noise and their pairing combinations. The intercorrelations are studied between acoustic dimensions and perceived soundscapes, annoyance, arousal, quality of life, and health indices. In addition, several measures at the contextual, social, and psychological levels—sometimes referred to as nonacoustical factors—are considered in the analyses. The results shed light on the validity of the used instruments to measure and predict the perceived soundscape and open a perspective to a better understanding of the major determinants of adverse, protective, and restoring effects of the various acoustic environments.

Chapter 6 introduces binaural measurement technology and psycho-acoustics, which are considered necessary components of a comprehensive soundscape investigation. While in numerous soundwalks the acquisition of soundscape-relevant sounds has been performed with binaural equipment and the achieved recordings used for subsequent analyses, this chapter presents the boundary conditions for applying this equipment regarding aspects such as measurement location, height, time, duration, orientation, and the use of a windscreen. The measurement protocols, which must consider measurement, environmental, and landscape-related aspects, are also discussed. Based on soundwalk investigations, the advantages and limitations of detailed psychoacoustic analysis of acoustic environments in the context of soundscape applications are outlined.

Chapter 7 presents some new mapping techniques that are useful tools to aid the design and planning process. This includes sound mapping, which shows sound-level distribution considering more source types, both positive and negative, than just traffic noise: soundscape mapping based on human perception of sound sources, soundscape perception mapping developed using artificial neural networks, psychoacoustic mapping and mind mapping, and mapping of noticed sounds.

Chapter 8 explores approaches to urban soundscape management, planning, and design, which can be used advantageously in new urban planning and development and in new projects, such as new transportation facilities and new shopping areas to improve the sound quality of urban areas, creating "quiet" and providing restoration. Following a discussion on outdoor soundscapes, noise control and soundscape management are compared, and then the chapter focuses on soundscape planning and design methods, allowing a deeper analysis of the future physical and sociological phenomena, which brings together the population and their expectations, and allows the designers to go beyond the classical noise control procedures and criteria. Urban designers and planners can then use the approach as a tool to provide added value to new urban projects that can be related to higher-quality or quieter areas, such as urban parks, squares, or meeting places that provide quiet and restoration.

Chapter 9 explores soundscape as part of the cultural and natural heritage. The implications that the soundscape (as a resource) has in parallel with the visualscape on the best conservation and restoration of cultural and natural heritage are discussed. Techniques for the recognition of the soundscape as a "trademark" of the cultural tangible and intangible properties are implemented. Improvement of tools such as management plans suggested by the United Nations Educational, Scientific and Cultural Organization (UNESCO) are suggested, and several applications are presented. So far the preservation guidelines for cultural and natural heritage (buildings, squares, and landscape) consider a unique element of any modification or restoration of the visual sensation. Nevertheless, the way people perceive and enjoy cultural and natural heritage is multisensorial, and the soundscape has intangible, intrinsic, and irreplaceable value, which contributes to its outstanding universal value, making it unique, recognizable by the community, and attractive for tourists—it can be as important as the visualscape.

Chapter 10 outlines exemplary case studies from which conclusions can be listed and discussed and guidance on good practice can be drawn. This chapter includes a selection of practical examples illustrates key issues in soundscape intervention. These include projects designed following a specific soundscape approach, for example, Nauener Platz, Berlin, and projects in which good soundscape has been achieved through more implicit or intuitive means, for example, via waterscapes in Sheffield, UK; via arts and engagement in Brighton and Hove, UK. The chapter also includes examples of soundscape practice in policy, for example, soundscape management in the U.S. national parks. The chapter reviews a range of ways in which soundscapes can be modified, for example, planning, architecture, engineering, and landscape design. The chapter draws out some strands for encouraging better soundscape practice among relevant professionals, including potential practice-oriented research and partnership projects.

While each chapter has a lead author, there are generally a number of authors collaborating on each of the chapters, mainly from the COST Action supporters. The contribution of all the participants of this Action and the support from the COST programme and the COST office are gratefully acknowledged. Chapter 3 is based on conversations and exchange with a variety of people about the topic, and special thanks go to Andre Fiebig, Elise van Kempen, HannekeKruize, and Mari Murel. For Chapter 7, the authors are indebted to Jiang Liu, Lei Yu, Yiying Hao, Bo Wang, and Wei Yang for their contributions to the maps, and Mohammed Boubezari and J. Luis Bento Coelho for the useful discussions. The author of Chapter 8 greatly acknowledges the inestimable discussions held with A. Lex Brown and with Gary W. Siebein. The authors of Chapter 9 thank the Municipality of Nola and the local organizations that allowed the audio and video recordings as well as the administration of the questionnaires during the folk festival. Heartfelt thanks to Saverio Carillo for the invaluable support provided in the historical research. For their invaluable support for the work of the Noise Abatement Society and the Sounding Brighton applied soundscape series of projects, the authors of Chapter 10 thank, in addition to those whose work has already been attributed within the chapter, the following individuals and organisations for whom and which we are extremely grateful and without whose support the work could not have been achieved: Gloria Elliot, Chief Executive, and the Trustees of the Noise Abatement Society; from Brighton and Hove City Council: Matt Easteal, Simon Bannister, Linda Beanlands, Donna Close (now at the University of Brighton), Jim Mayor, Alan Buck, and Council Leader then Jason Kitcat; Caroline Brennan and the members of the Brighton and Old Town Local Action Team, and the Brighton and Hove Local Action Team Chairs; members of the EU COST Action TD0804 on Soundscape; ISO TC 43 / SC 1 / Working Group 54 on Soundscape Convenyor Dr Osten Axelssön; from and advisors to the UK's Defra and British Standards Institute's noise technical committees: Bernard Berry, Stephen Turner, Colin Grimwood, and Phil Dunbaven; Brighton & Hove Arts Commission, Arts Council England, Rockwool UK, and the Swedish Research Council Formas.

Jian Kang and Brigitte Schulte-Fortkamp

Contributors

Tjeerd Andringa
University of Groningen
Groningen, The Netherlands

Itziar Aspuru
TECNALIA Research
    and Innovation
Bilbao, Spain

Dick Botteldooren
Acoustics Research Group
Ghent University
Ghent, Belgium

Giovanni Brambilla
CNR-Institute of Acoustics
    and Sensors "O.M. Corbino"
Rome, Italy

A. Lex Brown
Urban Research Program
Griffith School of Environment,
    Griffith University Nathan
Brisbane, Queensland, Australia

J. Luis Bento Coelho
Instituto Superior Técnico
University of Lisboa
Lisbon, Portugal

Maria Di Gabriele
Department of Architecture
    and Industrial Design
Second University of Naples
Aversa, Italy

Max Dixon
Independent Consultant
London, United Kingdom

Danièle Dubois
CNRS (LAM)
(Musical/Acoustic Laboratory)
University of Paris, France

André Fiebig
HEAD acoustics GmbH
Herzogenrath, Germany

Klaus Genuit
HEAD acoustics GmbH
Herzogenrath, Germany

Truls Gjestland
SINTEFICT
Trondheim, Norway

Mike Goldsmith
Acoustic Consultant
London, United Kingdom

Catherine Guastavino
McGill University
Montreal, Canada

Jian Kang
Acoustics Group School
   of Architecture
University of Sheffield
Sheffield, United Kingdom

Ronny Klæboe
Institute of Transport Economics
Oslo, Norway

Catherine Lavandier
Université de Cergy-Pontoise
Brighton and Hove
France

Lisa Lavia
Noise Abatement Society
Brighton and Hove
United Kingdom

Peter Lercher
Department of Hygiene,
   Microbiology and
   Social Medicine
Division of Social Medicine
Medical University of Innsbruck
   (MUI)
Innsbruck, Austria

Luigi Maffei
Department of Architecture
   and Industrial Design
Second University of Naples
Aversa, Italy

Mats Nilsson
Department of Psychology
Stockholm University
Stockholm, Sweden

Anna Preis
Institute of Acoustics
Adam Mickiewicz University
   of Poland

Brigitte Schulte-Fortkamp
Institute of Fluid Mechanic
   and Engineering Acoustics
Technical University of Berlin
Berlin, Germany

Irene van Kamp
National Institute for Public Health
   and the Environment
Bilthoven, The Netherlands

Eike von Lindern
Institute for Housing
   and Urban Research
Uppsala, Sweden

Harry J. Witchel
Brighton and Sussex Medical School
Brighton and Hove
United Kingdom

Chapter 1

Acoustic Environments and Soundscapes

A. Lex Brown,[1] Truls Gjestland,[2] and Danièle Dubois[3]

[1]Urban Research Program, Griffith School of Environment,
Nathan, Griffith University, Nathan, Australia

[2]SINTEF ICT, Trondheim, Norway

[3]CNRS (LAM) (Musical Acoustic Laboratory) University of Paris, Paris, France

CONTENTS

1.1 ACOUSTIC ENVIRONMENT

Sound surrounds and envelops us, whether we are indoors or out, at work or at play, in cities or in the country. We hear voices, vehicles, birds, wind in trees, machinery, footsteps, raindrops, telephones, the hum and beeps of our electronics, dogs barking, and sometimes the sound of blood moving through our bodies. Sound, through speech, is still the medium of much of our communication with others, despite the ubiquity of message texts and emails. Sound is always present, and our ears are always switched on, even when we are asleep. We share an acoustic environment with any who occupy the same indoor or outdoor space that we do. This book is primarily about the acoustic environments that we experience (when not cocooned in the private acoustic space of headphones and recorded sound) in all the places of our cities, suburbs and villages, countryside and natural areas. While much of the emphasis in this book is on sound in outdoor space, many of the observations, principles, and approaches will also apply to hybrid indoor–outdoor spaces—malls and markets, transport stations, sports stadia, museums, and the balconies of our own dwellings—and to indoor spaces in hospitals, educational institutions, restaurants, and even our dwellings. We have a particular focus in this book on the planning, design, and management of the environment of these areas, and human experience of them. Acoustic installations, and the recording of outdoor sound, as deliberate works of art, are not within the scope of this book.

The *acoustic environment* of a place or space is the sound from all sources that could be heard by someone in that place. This acoustic environment is shaped by all the different sound sources that are present and also by modification of the sounds as they propagate along their paths from the sources to the receiver. This modification includes reflection and absorption of sound by any surfaces present, including those of the ground surface and, to some extent, from vegetation, and both attenuation and refraction of sound along the source–receiver path as it passes through the atmosphere. In outdoor areas, there will often be only one reflecting surface—the ground surface—as the sound travels from the source to the receiver. In urban areas, there additionally will be reflections off walls and buildings, and in highly built-up areas, the acoustic environment becomes more like the semireverberant space that we experience indoors, with multiple reflections of sound as it travels from the sources to the receptor. Indoor acoustic environments are strongly shaped by these reflections; outdoor acoustic environments, particularly over longer propagation paths of hundreds of metres or more, are more shaped by the absorption of sound by the atmosphere and by refractive bending of the path of sound between sources and the receiver. The acoustic environment of any place thus depends on the sources present, the location of the receiver, and the propagation conditions along the path. Each of these may vary from instant to instant, from day to night, and from season to season. The *acoustic environment* (Maher 2004) of a place has also been called the *sonic environment* (Schafer 1997) or the *sound environment* (Yang and Kang 2005), and people have been described as being in an *environment of sound* (Truax 1999) or in *aural space* (Schulte-Fortkamp and Lercher 2003). Others might consider the acoustic environment to be the *ambient sound* of a place.

The study of acoustic environments of different places is of interest to many different fields of study and practice. For example, the acoustic environment of urban areas is of interest in urban planning and housing design and in landscape planning and management. The acoustic environment of natural areas is of interest to the fields of wilderness management, wildlife parks, and recreation. All acoustic environments are of interest to those charged with environmental noise management and to specialists in sound quality, human acoustic comfort, and music. It will be seen throughout this book that the study, analysis, and management of the acoustic environment involves diverse fields of practice, diverse approaches, and diverse disciplinary interests.

1.2 SOUND SOURCES IN THE ACOUSTIC ENVIRONMENT

Much of the distinctiveness of any acoustic environment lies in the presence or absence of particular sound sources, and it can be useful to identify each of the sources present and classify them. One proposed system has been to categorize all sounds as being of *biophonic, geophonic,* or *anthrophonic* origins (Gage et al. 2004). Biophonic sources are biological in origin, such as insect and bird sources. Geophonic sources have nonbiological origins, being generated by physical processes such as wind, rain, and thunder, earthquakes and volcanic activities, flowing water and ocean waves. Sources of sound induced by human activities are classified as anthrophonic. While some authors have used this system in nonurban areas (e.g., Pijanowski et al. 2011), the classification has insufficient resolution to be useful in urban areas given the very wide range of sources associated with human activity. More differentiation is required, and Brown et al. (2011) have developed a schema (Figure 1.1) that covers all sound sources in any acoustic environment. Their purpose was to provide a basis for standardization in the reporting of sources across different studies. The two systems can be integrated, as shown in Figure 1.1, where the anthrophonic category includes a large and comprehensive list of sound sources generated by human activity.

Figure 1.1 first categorizes places as outdoors or indoors, and then outdoor places as belonging to urban, rural, wilderness, or underwater domains. While human experience of the underwater acoustic environment may be limited, its sounds are increasingly being revealed through underwater recordings or in real time using transducers, for example, for whale-watching activities. One can thus refer, for example, to the *acoustic environment of a wilderness place* or the *acoustic environment of an urban place.* Having identified a place domain, Figure 1.1 then identifies *categories* of sound source, and then the sound sources themselves, that *could* be present in that place. For simplicity and comprehensiveness, the categories are the same for each domain, though some sources would be unlikely to be found within particular domains. This categorization can be regarded as a practical taxonomy of sound sources that can be used as a conceptual framework for reporting and analysis, and that will facilitate information retrieval and valid comparison of sound sources across different places and different studies. It is intended to be universal—able to be applied in all types of acoustic environments and places and thus portable across different studies. To the extent possible, it sets out to identify categories of sound sources that are mutually exclusive, unambiguous, and comprehensively inclusive of all possible sources. For example, a wilderness acoustic environment will consist largely of sounds not generated by human activity—the sounds of nature—but there could also be some human-generated sounds: aircraft, the speech or laughter of recreationists,

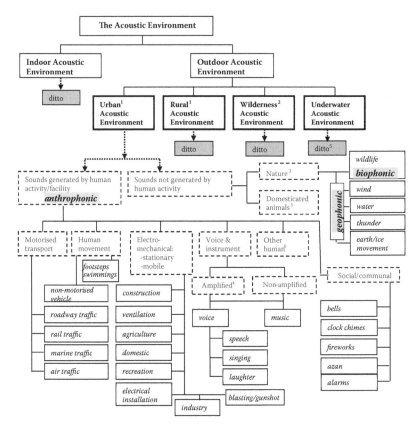

Figure 1.1 A classification scheme for categorizing sound sources in any acoustic environment that can be used to standardize sound source reporting across different studies. For each domain shown in bold boxes (urban, rural, wilderness, and underwater), there is an identical set of *categories of sound sources* shown in the dashed boxes, together with examples of *sound sources* in the different categories (solid boxes). (Modified from Brown et al., *Applied Acoustics* 72(6): 387–392, to include the anthrophonic, geophonic, and biophonic classifiers used by others.)

[1] The urban–rural divide will not always be distinct.

[2] The wilderness category includes national parks, undeveloped natural and coastal zones, large recreation areas, and so forth, and the wilderness–rural divide will not always be clear-cut.

[3] While "nature" and "domesticated animals" sources are shown as being "not generated by human activity," there are many areas of overlap—for example, the sounds of running water in constructed water features or the sounds of wind on buildings. Domesticated animal sounds will generally be from animals associated with a human activity/facility.

[4] Recording, replay, and amplification may occur for any type of sound—as, for example, in installations playing nature/wildlife sounds.

[5] Because of the different acoustic impedances in air and water, many of the terrestrial sound sources of the figure would not normally be observed under water, but overall the same classification system is still applicable.

[6] Coughing, for example.

and perhaps the amplified speech from a ranger's radio. In the courtyard of a housing estate, sounds generated by nature may be incidental, and those generated by human activity will be present. In some places, certain sounds of human activity, say, footsteps, may be present, with only infrequent sound from roadway traffic, but in another, roadway traffic may constitute the only sound source that can be heard. In each of these examples, a systematic taxonomy of sources encourages unambiguous and value-free description of them, and provides a common terminology.

The nomenclature of the categories and sound sources has also been carefully chosen to avoid imprecise, or polysemic, labels and descriptors—often these have not been applied in a uniform way to the sources of sound in different acoustic environments. They have also been chosen to avoid imputing value judgements to the source within any specific context. This circumvents, for example, a particular sound source being described as a *background* sound in one place but a *foreground* sound in another, *intrusive* in one place but *acceptable* in another.

Here we have considered only the classification of sources of sounds. In other chapters of this book, we discuss the complexities of measurement of the acoustic properties of the individual sources that can be described by acoustical parameters such as sound level, spectrum, and temporal pattern, and procedures for the measurement and description of the overall acoustic environment (Gjestland 2012; Kihlman and Kropp 2001).

1.3 SOUNDSCAPE

1.3.1 Soundscape as a Perceptual Construct

Most authors suggest the *soundscape* of a place is a person's perceptual construct of the acoustic environment of that place (Porteous and Mastin 1985; Truax 1999; Finegold and Hiramatsu 2003; Gage et al. 2004; Brown and Muhar 2004; Yang and Kang 2005; Dubois et al. 2006; Kang 2006). That place will be a physical, often outdoor, area (or space or location) that will have certain visual and other properties as part of its human-made or natural environment. The place will also be where people might live or occasionally spend time and where they undertake active or passive activities, and in which they interact with its environment and with others. A person undertaking some activity in this place experiences auditory stimulation from the acoustic environment and interprets the auditory sensations arising from this stimulus. A process framework of perception and interpretation of an acoustic environment is illustrated in block form in Figure 1.2. The resulting perceptual construct of the acoustic environment can be termed the *soundscape*. The soundscape has the potential, within that particular context, to evoke responses in the individual and result in outcomes that can be attributed to it. The perception of the soundscape and the responses

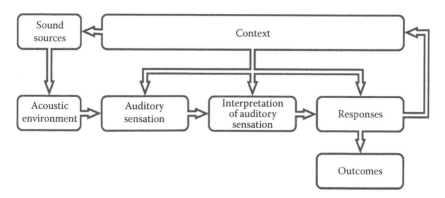

Figure 1.2 A conceptual diagram of the elements of the process by which an individual's perceptual construction of a soundscape occurs within an acoustic environment. The context (physical as well as previous knowledge and experience of the individual) in which this process occurs critically influences the perception of the soundscape, as it does the human response to it, and outcomes arising from the experience. (From ISO/FDIS 12913-1. Acoustics—Soundscape—Part 1: Definition and Conceptual Framework, April 2014.)

and outcomes are highly dependent on the context, and relevant contextual factors are examined further in Section 1.6.

1.3.2 Responses and Outcomes

Some examples of the many possible responses that can be attributed to the experience of an acoustic environment include pleasure, acoustic comfort (Kang 2006), excitement, or fear. Example outcomes might include place attachment, a sense of harmony, restoration of well-being, or perhaps appreciation of nature through bird calls or sounds of other species heard in a natural area. In different places and in different contexts, human preferences in terms of response to the acoustic environment, or outcomes attributable to it, may differ markedly, or be multidimensional. For example, the soundscape of a place might be preferred on the basis that it is peaceful, or tranquil, or promotes well-being. Equally, in a different place or context, a soundscape might be preferred because it is lively, or varied, or creates a sense of excitement. Or preference may be for a soundscape that provides information, clarity, or conveys safety. In yet another place or context, preference for a soundscape may relate to its unique cultural or natural characteristics—a place with soundmarks (Schafer 1977). Those working in particular fields may have a very clear mandate regarding particular response and outcomes in which they are interested (say, in national parks, recreation, or wilderness areas), but these are likely to be different than the responses or outcomes of interest to those working in other fields, say, urban open space or housing design. Given such diversity in contexts in which acoustic environments of particular places may be experienced, and diversity in potential responses and outcomes from this experience, more research is required to identify all the sets of responses and outcomes that could be of interest in soundscape studies, to develop a typology of these, and to search for underlying structures and association within them.

The outcomes illustrated above can be considered *direct outcomes*. Direct outcomes depend, to a large extent, on people being aware of the acoustic environment—and consciously attributing the outcome directly to the perceived soundscape. However, the acoustic environment of a place may also enable certain responses/outcomes without people consciously dissecting why it is that the environment of a place provides so well for that activity. For example, people may know that a park is a good one in which to play with children, relax or meditate, meet with people, communicate, or undertake other activities. They seek to achieve these outcomes in places—facilitated by the acoustic environment, along with other dimensions of the place—but not necessarily with conscious attention to the acoustic environment itself. Attempts to measure soundscape preference in this situation can pose a significant methodological problem, introducing an *experimenter effect* where measurement of preference requires first drawing a respondent's attention to something on which he or she may have never consciously reflected. Assessment of soundscape preference should recognize the existence of both direct outcomes and *indirect* or *enabled outcomes*.

1.3.3 Other Uses of the *Soundscape* Term

There is not universal agreement among authors that the soundscape is a human perceptual construct. Some prefer to use the term as a synonym for the physical *acoustic environment*, that is, "the collection of sounds in a place" or (Oliveros 2005) "the sound variations in space and time . . . of the built-up city and its different sound sources" (Kihlman and Kropp 2001). As long as such equivocal usage of the term *soundscape* does not introduce ambiguity or confusion in communication, we can be relaxed about such use. Those likely to think of the soundscape as the collection of all sounds in a place will likely include planners, designers, laypersons, and even those primarily interested in management of the acoustic environment through environmental noise control. Pijanowski et al. (2011) refer to the soundscape as "all sounds . . . emanating from a given landscape to create unique acoustical patterns across a variety of spatial and temporal scales," but their focus, while still including human-generated sounds, tends primarily to be on the acoustical patterns created by biological systems, and their field of study is referred to as *soundscape ecology*.

While the focus of this book, mainly grounded within the acoustic community, is on physical places with the sound sources of the acoustic environment heard in real time, the soundscape as a perceptual construct can apply to representations of the acoustic environment in memory (Ge and Hokao 2003), to the "assumed acoustic environment" of a historic place or event (e.g., Forum Romanum in ancient Rome or the Civil War battle at Gettysburg), or even to abstracted acoustic environments such as musical compositions (Schafer 1977). Even more broadly, the soundscape terminology has also encompassed, for example, the recording of the sounds of nature, the creation of compositions based on or of natural sounds, studies of the sounds heard in villages and rural environments, documentation of disappearing sounds, analysis of the way acoustic environments have been described in history and in literature, analysis and description of all types of acoustic environments, and the creation of artistic sound installations (Torrigoe 2003). While we acknowledge all of these different uses of the term, the focus in this book is on the soundscape as the human perceptual construct of the acoustic environment of any place, and on the planning, management, manipulation, or design of the acoustic environment of a place to change human response to it.

1.4 SOUNDSCAPE–LANDSCAPE ANALOGY

Various authors have drawn a useful analogy between the *soundscape* and *landscape* terms (Porteous and Mastin 1985; Brown and Muhar 2004; Maher 2004; Oliveros 2005; Pijanowski et al. 2011). *Landscape* is regarded as both a perceptual construct and a physical phenomenon (Appleton 1996; Benson and Roe 2000). The European Landscape Convention Agreement (Council of Europe 2000) provides a good definitional basis, defining landscape as "an area, as perceived by people, whose character is the result of the action and interaction of natural and/or human factors." Substituting *place* for *area* because of the high spatial variability of the acoustic environment over any of the types of outdoor areas in which we are likely to be interested, the useful parallel definition of soundscape is "the acoustic environment of a place, as perceived by people, whose character is the result of the action and interaction of natural and/or human factors."

The interpretations that can be placed on the term *soundscape* can be as diverse as the different interpretations people already have of its namesake *landscape*—for example, the latter can include landscape as geographical form, landscape as a system of physical components, landscape as both determinant and reflection of culture (painting, literature, and music), landscape as a place for recreational activity, and landscape in the design activity of landscape planning or architecture. By analogy, a similar diversity applies to interpretations/applications of the *soundscape* term.

The European Landscape Convention goes on to define *landscape policy*, *landscape planning*, and *landscape management*. Again by analogy, this usefully results in *soundscape policy* being "the expression by the competent public authorities of general principles, strategies and guidelines aimed at the protection, management and planning of soundscapes." It is similar for the self-explanatory expressions *soundscape management* and *soundscape planning*, terms that, like *soundscape design* (Truax 1998), underlie the emphasis in this book on the consideration of the acoustic environment in the planning and management of outdoor space.

This section has focused only on the analogy that exists between the terms *landscape* and *soundscape*. However, there is evidence that there is real interaction between the acoustic attributes of a place and the visual attributes of a place in terms of human environmental perception and experience (Carles et al. 1999; Ge and Hokao 2003).

1.5 *SOUNDSCAPE* IN DIFFERENT LANGUAGES

Language differences are involved in some of these different interpretations of the word *soundscape*. To shed some light on this subtlety and complexity, a small survey was conducted of various researchers involved with the COST Action TD0804, drawn from different countries and cultures (and from very different scientific disciplines). First, we sought translations for the word *soundscape*, successfully obtaining these from 14 European languages belonging to different linguistic families, and from the United States, Korea, and Japan. These words reflect the difficulties in translations along with the properties and constraints specific to the languages themselves. Norwegian, Swedish, and Dutch, languages related to English, allow one direct translation by copying the morphological construction of English *sound + scape*, respectively, *Lyd + landskap* in Norwegian, *Ljud + landskap* in Swedish, and *geluids + landschap* in Dutch. A similar construction can be found in Finnish, *ääni + maisema*, and also in Greek, ηχο + τοπίο. Roman languages, however, seem to have an *adjective + noun* construction, and there are syntactic constructions imposing choices on both the adjective and the noun. If there is a clear agreement on the adjectives *sonore*, *sonoeo*, and *sonore*, along with *acoustic*, there are noticeable differences in translating the -*scape* part of the word: *paysage*, *environnement*, and *milieu* (French); *paesaggio*, *clima*, *ambiente*, *scenario*, and *impronta* (Italian); *paisaje* and *ambiente* (Spanish); *krajobraz*, *pejzaż*, and *sfera* (Polish); *peyzaj* and *alan* (Turkish); and so on.

These differences may not be trivial and indicate that the concept *soundscape* may be perceived or understood as the acoustic correspondence to either *landscape*, that is, shifting from the visual domain to the acoustic one as a two-dimensional (2D) conception of perception where humans are "observers" facing an (common and shared) objective world (as identified and measured by physical sciences), or *environment*, which refers to a three-dimensional (3D) conceptualization or, when worded as *ambiance*, a more immersive and global multisensory experience of the world by humans, where the acoustic stimulations are just part of the whole setting.

When asked to answer the question "Could you list the English words (as many as you wish) that you (would) use to get (bibliographic) information" the answers (worded in English) show, beyond a large variability (or a lack of consensual wording),

- A frequent reference to the *environment* through the noun itself (mainly *acoustic environment, sound environment, sonic environment*, etc.) or through an adjectival form (*environmental*), qualifying *sound* as well as *noise*, along with names such as *urban* and *natural*, referring to more specific environments.
- *Sound* (either as noun or as a qualifier in a compound noun) occurs more frequently than *noise*—not unexpected given the neologism chosen was *soundscape* and not *noisescape*.
- Even if *acoustic* and *acoustical* remain more frequent, some mention of *auditory, aural*, and *perception*, as well as the word *quality* (*sound quality* and *quality of life*).

1.6 ROLE OF CONTEXT

Figure 1.2 suggested that *context*, as a generic term that includes all other nonacoustic components of the place (even including people's experience and memory), plays a major role in people's perceptual construction of soundscapes. But that diagram was short on detail as to what those contextual components might be. Herranz-Pascual et al. (2010) have addressed this problem and situate environmental experience of sounds in public places firmly within an environmental people–activity–place framework. They group contexts that they suggest will be relevant in four clusters: person, place, person–place interaction, and activity (Figure 1.3). Each of the elements in these clusters potentially influences a person's experience of the acoustic environment there. Each place has its own particular landscape or built form, and aesthetic, physical (including acoustic) and social/cultural attributes, and these will change with time of day, season, or weather. Each person brings with him or her to that place his or her own demographics, perceptions, lifestyle, culture, networks, attitudes, and preferences, in terms of immediate matters, such as motivation for being in a place and undertaking an activity, and also longer-term attributes individuals may carry with them. The person and place interact through previous experience of that place, its familiarity and identity, and information about and expectations of that place. The person may undertake active or passive activities there, alone or in social interaction, and will have certain activity needs and opportunities. Changing any one of the elements within this model, even if all other elements remain constant, could significantly change the person's environmental experience of that place at that point in time, and hence his or her perceptual construction of the soundscape of that place.

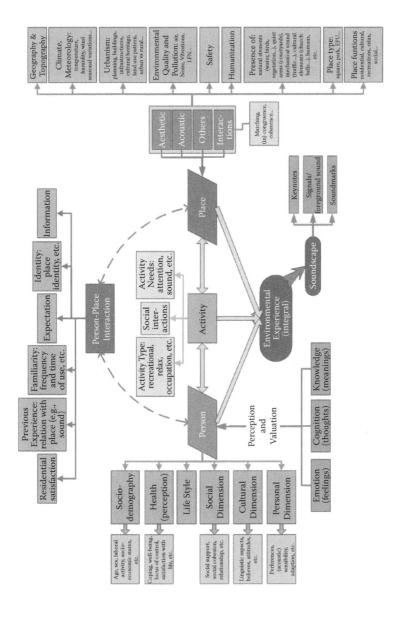

Figure 1.3 Conceptual model of various contexts of a person, place, person–place interaction, and activity relevant to that person's perceptual construction of the soundscape of that place. (From Herranz-Pascual, K., et al., Proposed conceptual model of environmental experience as framework to study the soundscape, in *Inter Noise 2010: Noise and Sustainability*, Lisbon, Portugal, 2010.)

This means that any study of soundscapes, or measurement of people's soundscape preference, must be cognizant of the role of context and aware of the variability in soundscape preference or outcomes that arise from any change in these. What is preferred in one place will be different than what is preferred in another. This variability will be between people, and even within one person in one place, at different times of a day, a week, or a season.

There is nothing either unusual or capricious about such human variability. There will be variation between groups of people of dissimilar age (different types and intensity of music, for example), social status, and so forth. Despite such diversity, it is suggested that there will often tend to be more agreement than disagreement between people regarding preferred soundscapes of any place (Stack et al. 2010). In any case, planners, designers, and managers deal every day with the complexity and variability of community response (Brown 2012).

There is increasing evidence that it is the congruence between the different elements of a place that is important to human preference (Brambilla and Maffei 2006), and perhaps between these elements and expectations of the place.

1.7 COMPARING SOUNDSCAPE AND ENVIRONMENTAL NOISE MANAGEMENT APPROACHES

The acoustic environments of outdoor space, here of interest in terms of soundscapes, have, to date, largely been considered the province of those charged with environmental noise management. Truax (1998) distinguishes between these two approaches to the outdoor acoustic environment: environmental noise management is the *traditional, objective energy-based model of the acoustic environment*; the soundscape approach is a *subjective listener-centred model*. It is useful to examine how the two approaches differ, and how they extend and complement each other. The different foci of the two approaches are shown in Table 1.1.

Sound is conceived as a waste product in the environmental noise field—a waste to be reduced and managed. Such noise reduction or management may be at the source of the noise, in the propagation path, or at the receiver itself. Brown (2010) suggests that, in contrast, "the soundscape field regards

Table 1.1 Different Foci of Environmental Noise and Soundscape Approaches

Environmental Noise Management Approach	Soundscape Approach
Sound managed as *a waste*	Sound perceived as *a resource*
Focus is on *sounds of discomfort*	Focus is on *sounds of preference*

Source: Brown, A. L., *Noise Control Engineering Journal* 58: 493–500, 2010.

sound largely as a resource—with the same management intent as in other scarce resources such as water, air and soil: rational utilization, and protection and enhancement where appropriate. Resource management has a particular focus on the usefulness of a resource to humans and its contribution to the quality of life for both present and future generations."

Another essential distinction between the two fields is the human outcome of interest. By and large, environmental noise deals with adverse outcomes for people—or *sounds of discomfort* (Augoyard 1998): any or all of the effects of sleep disturbance, annoyance, adverse physiological effects, interruption to communication, and so on. The focus in soundscape studies is mostly on *sounds of preference*. Preference is considered within building acoustics in terms of preferred ambient levels for rooms, or preferred reverberation time in halls for speech and music, and for products in terms of their sound quality. Preference has to date had little use in environmental acoustics, though Genuit (2002) has suggested that sound quality concepts should also have environmental and soundscape application. It is sometimes suggested that the fundamental distinction between the soundscape field and the environmental noise field is its focus on human perception, but much work in environmental noise is also perceptually based, as in the measurement of annoyance. The real distinction between the two fields is the different human outcomes of interest.

1.8 SUMMARY

In the COST Action Soundscape of European Cities and Landscapes, the focus has been on improving or creating the acoustic environment of places so that the soundscape enhances human enjoyment. The interest is primarily, but not exclusively, in outdoor areas such as streets and squares, city parks, gardens, natural areas, or wilderness, but many of the observations, principles, and approaches in this book will also apply to hybrid outdoor–indoor spaces, such as malls and markets, transport terminals, sports arenas, and similar.

In this chapter, we have introduced the acoustic environment of such places, and a classification scheme for categorizing sound sources that will be found in any acoustic environment where the soundscape may be of interest. The categorization can be used to standardize sound source reporting across different places, and has been designed largely to be independent of place and free of value judgements or interpretations of different sources.

A *soundscape* is a person's perceptual construct of the acoustic environment of a place. The soundscape has the potential, within that particular context, to evoke responses in the individual and result in outcomes that can be attributed to it. It is human response to the soundscape and human preference for particular responses and outcomes that are of particular interest in this book.

Perception of the soundscape of a place and human responses and outcomes are highly dependent on context. The person and place interact through previous experience of the place, its familiarity and identity, information about and expectations of the place. Changing any one of the contextual elements within this *person–place–activity* model, even if all other elements remained constant, could significantly change the person's environmental experience of a place, and hence his or her perception of the soundscape. The experience of an acoustic environment, the soundscape, can therefore be different even if all physical parameters, acoustical and others, remain constant. For example, a particular acoustic environment may be considered good or relaxing if a person is seeking solitude or quiet, but neutral or uneventful if a person is seeking excitement.

In this book, we are unequivocal in describing the soundscape as a human perceptual construct, but such usage is not universal, with some using the term as a synonym for the acoustic environment, that is, the collection of sounds in a place. Those likely to think of the soundscape as the collection of all sounds in a place may include planners, designers, laypersons, and even those primarily interested in environmental noise management. As long as this does not introduce ambiguity or confusion in communication, or inhibit management action, we can be relaxed about such alternative interpretations of the soundscape.

There is a useful analogy between the terms *soundscape* and *landscape*. Borrowing from the definition of *landscape* in the European Landscape Convention Agreement, a *soundscape* can be defined as "the acoustic environment of a place, as perceived by people, whose character is the result of the action and interaction of natural and/or human factors." Extending the analogy to other terms in the convention agreement, *soundscape policy* is "the expression by the competent public authorities of general principles, strategies and guidelines aimed at the protection, management and planning of soundscapes," and similarly for the self-explanatory expressions *soundscape management* and *soundscape planning*. The latter usefully reflect the emphasis in this book on the planning and management of the acoustic environment of outdoor space.

The soundscape field intersects with the established field of environmental noise management—both are concerned with the outdoor acoustic environment. There are overlaps, but there are also distinct differences. In the environmental noise field, sound is conceived as a waste product. In contrast, "the soundscape field regards sound largely as a resource— with the same management intent as in other scarce resources such as water, air and soil: rational utilization, and protection and enhancement where appropriate. Resource management has a particular focus on the usefulness of a resource to humans and its contribution to the quality of life for both present and future generations." Soundscapes focus on the management of the acoustic environment as a resource that can contribute to human quality of life.

REFERENCES

Appleton, J. 1996. *The Experience of Landscape* (rev. ed.). New York: Wiley.

Augoyard, J.-F. 1998. The cricket effect. Which tools for the research on sonic urban ambiences? Papers presented at the "Stockholm, Hey Listen!" Conference, Royal Swedish Academy of Music.

Benson, J. F., and Roe, M. H. 2000. The scale and scope of landscape and sustainability. In *Landscape and Sustainability*, ed. J. F. Benson and M. H. Roe. London: Spon Press.

Brambilla, G., and Maffei, L. 2006. Responses to noise in urban parks and in rural quiet areas. *Acta Acustica united with Acustica* 92(6): 881–886.

Brown, A. L. 2010. Soundscapes and environmental noise management. *Noise Control Engineering Journal* 58: 493–500.

Brown, A. L. 2012. A review of progress in soundscapes and an approach to soundscape planning. *International Journal of Acoustics and Vibration* 17(2): 73–81.

Brown, A. L., Kang, J., and Gjestland, T. 2011. Towards standardization in soundscape preference assessment. *Applied Acoustics* 72(6): 387–392.

Brown, A. L., and Muhar, A. 2004. An approach to the acoustic design of outdoor space. *Journal of Environmental Planning and Management* 47: 827–842.

Carles, J. L., Barrio, I. L., and de Lucio, J. V. 1999. Sound influence on landscape values. *Landscape and Urban Planning* 43: 191–200.

Council of Europe. 2000. European Landscape Convention. ETS No. 176. http://conventions.coe.int/Treaty/en/Treaties/html/176.htm.

Dubois, D., Guastavino, C., and Raimbault, M. A. 2006. Cognitive approach to urban soundscapes: Using verbal data to access everyday life auditory categories. *Acta Acustica united with Acustica* 92: 865–874.

Finegold, L., and Hiramatsu, K. 2003. Linking soundscapes with land use planning in community noise management policies. In *InterNoise 2003*, Seogwipo, South Korea.

Gage, S., Ummadi, P., Shortridge, A., Qi, J., and Jella, P. K. 2004. Using GIS to develop a network of acoustic environmental sensors. In *ESRI International Users Conference*, 15–28, San Diego, CA.

Ge, J., and Hokao, K. 2003. Research on the formation and design of soundscape of urban park: Case study of Saga prefecture forest park, Japan. In *International Symposium on City Planning*, Sapporo, Japan.

Genuit, K. 2002. Sound quality aspects for environmental noise. In *Inter Noise 2002*, Dearborn, MI.

Gjestland, T. 2012. Reporting physical parameters of soundscape studies. In *Acoustics 2012*, Nantes, France.

Herranz-Pascual, K., Aspuru, I., and García, I. 2010. Proposed conceptual model of environmental experience as framework to study the soundscape. In *Inter Noise 2010: Noise and Sustainability*, Lisbon, Portugal.

Kang, J. 2006. *Urban Sound Environment*. London: Taylor & Francis.

Kihlman, T., and Kropp, W. 2001. Soundscape support to health: A cross-disciplinary research programme. In *InterNoise 2001*, The Hague, The Netherlands.

Maher, R. C. 2004. Obtaining long-term soundscape inventories in the U.S. National Park System. White paper prepared for National Park Service Natural Sounds Program Office, Fort Collins, CO.

Oliveros, P. 2005. *Deep Listening: A Composer's Sound Practice.* iUniverse, Inc., 2005.

Pijanowski, B. C., Villanueva-Rivera, L., Dumyahn, S., Farina, A., Krause, B., Napoletano, B., Gage, S., and Pieretti, N. 2011. Soundscape ecology: The science of sound in the landscape. *BioScience* 61(3): 203–216.

Porteous, J. D., and Mastin, J. F. 1985. Soundscape. *Journal of Architectural and Planning Research* 2: 169–186.

Schafer, R. M. 1977. *The Tuning of the World.* New York: Alfred A. Knopf.

Schulte-Fortkamp, B., and Lercher, P. 2003. The importance of soundscape research for the assessment of noise annoyance at the level of the community. *Tecni Acustica*, Bilbao.

Stack, D. W., Newman, P., Manning, R. E., and Fristrup, K. M. 2010. Reducing visitor noise levels at Muir Woods National Monument using experimental management. *Journal of the Acoustical Society of America* 129(3): 1375–1380.

Torrigoe, K. 2003. Insights taken from three visited soundscapes in Japan. Proceedings of World Forum for Acoustic Energy. Australian Forum for Acoustic Ecology, Fairfield, Victoria, Australia.

Truax, B. 1998. Models and strategies for acoustic design. Paper presented at "Stockholm, Hey Listen!" Conference, Royal Swedish Academy of Music.

Truax, B. 1999. *Handbook for Acoustic Ecology* (2nd ed.). Cambridge Street Publishing. Burnaby, B.C. Canada.

Yang, W., and Kang, J. 2005. Soundscape and sound preferences in urban squares: A case study in Sheffield. *Journal of Urban Design* 10: 61–80.

Chapter 2

From Sonic Environment to Soundscape

Dick Botteldooren,[1] Tjeerd Andringa,[2] Itziar Aspuru,[3] A. Lex Brown,[4] Danièle Dubois,[5] Catherine Guastavino,[6] Jian Kang,[7] Catherine Lavandier,[8] Mats Nilsson,[9] Anna Preis,[10] and Brigitte Schulte-Fortkamp[11]

[1]Ghent University, Ghent, Belgium

[2]University of Groningen, Groningen, The Netherlands

[3]TECNALIA Research and Innovation, Bilbao, Spain

[4]Griffith University, Brisbane, Australia

[5]Pierre and Marie Curie University, Paris, France

[6]McGill University, Montréal, Canada

[7]University of Sheffield, Sheffield, United Kingdom

[8]Université de Cergy-Pontoise, France

[9]Stockholm University, Stockholm, Sweeden

[10]Adam Mickiewicz Univeristy, Poznań, Poland

[11]Technical University of Berlin, Berlin, Germany

CONTENTS

Chapter 2 focuses on creating deeper understanding of the relationship between the sonic or acoustic environment and soundscape. It attempts to relate the somewhat vague concept *soundscape* to findings from psychophysics, psychology, hearing system physiology, and auditory cognition. The term *soundscape* has been used by different communities of practice (e.g., acousticians, composers, architects, ecologists, and psychologists), giving rise to several definitions (see Chapter 1). A standardized definition may not be required, but it is useful to summarize generally accepted views on this concept:

- The soundscape is evoked by the physical sound environment—henceforth called the *sonic environment*—but it is not equal to it, and therefore cannot be measured using classical sound measurement equipment alone.
- The soundscape is formed within a context. This context is shaped by all sensory stimulations—of which visual observations are most important—and by the knowledge people have accumulated about the place, its use, its purpose, its cultural meaning, their own and others' motivations and purposes to be there, the associated activities, and so forth.
- The soundscape concept tends to be used mostly in relation to open outdoor places, but it also has applications for indoor settings, mainly

public, but also private. It always entails a sense of spaciousness. Environmental sounds intruding in private space result in effects following different mechanisms, with control as an important factor.

• The timescale related to soundscapes is in the order of minutes to hours. The quality of the soundscape in some parts of the living environment can nevertheless have long-term effects on the quality of life and health of the population (see Chapters 3 and 5).

The following sections discuss various aspects of the relationship between the sonic, or acoustic, environment of a place and the person experiencing the soundscape to finally construct a holistic model. After a brief discussion of listening styles, this chapter will focus on low-level auditory scene analysis that leads to the formation of auditory streams and objects. It will then continue by introducing the important role of auditory attention in selecting which of these auditory objects will be noticed. Attended sounds get meaning, so this important aspect of soundscape analysis and design will be the focus of the next section. Finally, appraisal and quality judgement will be addressed as the last step in the process. Once our current knowledge on each of these aspects has been thoroughly investigated, a more holistic view on the relationship between the sonic environment and soundscape will be discussed. The last section of this chapter will highlight how this knowledge can be applied in practice for measuring and designing soundscape. This last part could be seen as an introduction to the next chapters in this book.

2.1 LISTENING

Listening is a complex process that involves multileveled attention and higher cognitive functions, including memory, template matching, foregrounding (attentive listening), and backgrounding (holistic listening) (Truax, 2001). Some scholars group listening styles in everyday listening and musical listening, thereby focusing on an apparent difference between music, a sound that is produced for a purpose, and all other sounds (Gaver, 1993a, 1993b). Everyday listening in this terminology focuses on the sound source, musical listening of the sound itself.

2.1.1 Attentive, Analytic, Descriptive Listening: Most Popular in Soundscape Research

In investigations where persons are asked about their aural experience in a place, researchers found that these persons most often mention particular sounds by naming the source of these sounds (McAdams, 1993). One could conclude from this that attentive, analytic, descriptive listening is the most important listening style in relation to the soundscape experience.

This is, however, only partly true. Asking visitors of a place to describe their listening experience automatically triggers attentive and descriptive listening. In absence of the researcher, this listening style would only be important in those special cases where the intended activity includes a strong attention focus on the environment or when the sound is so prominent and salient that listening to it cannot be avoided. Even musical listening, although it focuses on the sound rather than on its sources, can still be regarded as a type of attentive listening.

2.1.2 Holistic Listening and Hearing: A More Hidden Contribution to the Soundscape Experience

One should not underestimate the potential role of holistic listening or even simply hearing as a mediator in creating mood and appraisal of the sonic environment. Sound not actively attended to, and thus pushed to a background, can still have meaning (see further).

To our knowledge, there is no research that directly relates soundscape listening to the cognitive effort it requires. However, as attention assigns more cognitive resources to a sensory input stream, it is reasonable to assume that attentive listening, and in particular analytic, descriptive listening, requires more cognitive effort and is, as a consequence, also slower. Holistic listening and backgrounding frees cognitive resources for other tasks that might be more relevant at this instant in time. Yet, holistic listening is also expected to be faster, and thus allows the organism to more quickly create a mental image of its environment and act correspondingly.

2.1.3 Different Listening Styles: All Part of the Same Experience

As the listening experience in a sonic environment evolves, the listener switches between different listening styles: from the more holistic listening in readiness, waiting for familiar or important sounds to emerge (expected or not), to listening in search, expecting particular sounds in a context, or even to narrative or story listening—musical listening could be seen as a specific example of this listening style—focusing attention on one particular sonic story within the multitude of sounds.

2.2 AUDITORY SCENE ANALYSIS

The sonic environment of interest in the context of soundscape consists of a multitude of individual sounds. One of the first tasks of the auditory system is to analyze this auditory scene and identify its building blocks, a process referred to as auditory scene analysis (ASA). ASA involves decomposing a complex mixture of incoming sounds, originating from different sources, into individual auditory streams, using different auditory, but also visual and other, cues (Bregman, 1994). Auditory streams are classically regarded as existing in a preattentive phase. Although this view is appealing because of its conceptual simplicity, recent findings suggest that attention also plays a role in the formation of auditory streams (Cusack et al., 2004; Shamma et al., 2011). Overall, it can be stated that the process of auditory scene analysis draws on low-level principles for segmentation and grouping, but is fine-tuned by selective attention (Fritz et al., 2007). Sound objects within the sonic environment are thus formed with the help of selective attention (attention mechanisms will be explored further in the following paragraphs). In relation to the discussion in the previous paragraph, this implies that even in holistic listening, stream and object formation occur, yet they may be less precise than during attentive, descriptive listening.

Scene analysis is partly multisensory, although the relative importance of the components of this scene analysis in vision and audition is different. In auditory scene analysis, temporal grouping at timescales from seconds to minutes or even hours is extremely important. Grouping at shorter timescales is more likely to occur—at least as a first estimate—during the preattentive phase. Spatial cues obtained through binaural hearing also help the stream segregation process. Binaural unmasking is known to be a key factor in targeted story listening within a masking background noise. With environmental sound and soundscape in mind and with the relative importance of different listening styles discussed above, binaural cues may be less significant for stream segregation in this listening context.

2.2.1 Auditory Scene Analysis and Familiarity with Sounds

Identification of auditory objects based on spectrotemporal features is a learned process where learning relies on co-occurence of these features. The importance of temporal coherence in auditory scene analysis and learning in humans has recently been confirmed on a neurological basis (Shamma et al., 2011). As such, the familiarity of the listener with a sound may contribute to the ability to distinguish this sound in a complex sonic environment. Prior experience could therefore even have an influence on this low-level preattentive ASA and lead to interindividual differences in perceiving the sonic environment and in the soundscape experience. This ASA skill is transferable between sounds, as the ability to group and identify features may be influenced by early sound experience of any style: language, music, or even the simplest sounds in the daily living environment. This could also lead to cultural differences, as typical language and musical sounds may differ across cultures.

2.2.2 Effects of the Complexity
of Auditory Scene Analysis
on Soundscape

In general, one could expect that a sonic environment where various auditory streams can easily be formed is appreciated as a high-fidelity soundscape. More complex situations that cannot easily be "read" by the average listener may be perceived as too complex and mentally stressing. Like in the attention restoration theory (ART) and the associated fascination, urban environments, with too complex stimulations, could be a source of attentional fatigue (Payne, 2013; Kaplan and Kaplan, 1989). If, on the contrary, ASA results in a single auditory stream, the sonic environment may be perceived as boring.

The ability to segregate sounds from complex mixtures differs between persons, as does the amount of cognitive resources needed for this task. Aging, for example, does not affect the ability for sequential streaming, but it has a pronounced effect on concurrent sound segregation (Snyder and Alain, 2007). These interindividual and age-related differences in even this lowest level of processing may lead to differences in the degree of complexity in the sonic environment that is desirable.

2.3 ROLE OF ATTENTION

Let us now focus on the attention mechanism in more detail. The role of selective attention is to allow part of the sensory input to be evaluated in the context of specific knowledge while preventing sensory signals from overloading the higher-level cognitive system. Overall perceptual load thus plays an important role in attention mechanisms (Lavie et al., 1994; Lavie, 2010). In situations where soundscape analysis and design are usually applied, the use of the place does not focus on communication between a performer and a group of listeners (such as a musical performance or theatre). Apart from the verbal communication they may be involved in, most users of the public space have little interest in listening for particular sounds, such as birds or insects. However, as the auditory system always stays alert, sounds within the sonic environment could draw attention. The proposed theoretical model foresees a two-stage mechanism to account for this: auditory stimuli may draw attention because of specific features they possess, but they don't necessarily get attended. This two-stage mechanism is supported by neuroscience: sounds with high saliency trigger early brain response (Escera et al., 1998), while inhibition of return (Prime et al., 2003) and voluntary attentiveness to sound determine whether a late response corresponding to actual attending is observed.

2.3.1 Saliency-Driven Attention

Identifying sound features that increase saliency (Kayser et al., 2005) and attract attention is an important aspect of the proposed soundscape theory. It is well known (Kayser et al., 2005) that spectral and temporal variations and modulations—sometimes referred to as ripple—increase saliency for human observers. However, the auditory brainstem, which is responsible for these specific sensitivities, has a much higher plasticity than originally thought. On the basis of this, one could expect a common basis for auditory saliency, but in addition, some specificity for different (groups of) people.

Saliency sound features have been implicitly used in earlier studies in other fields, such as noise annoyance and sound quality. Tonality (Hellman, 1982), rhythm, or periodicity and impulsiveness have been introduced to explain differences in annoyance caused by different (industrial) environmental sounds (ISO 1996-1, 2003). Likewise, sharpness, roughness, and relative approach (Genuit and Fiebig, 2006) describe features of the sound that attract attention. And, of course, the loudness of the sound itself is an important factor in its saliency. Event-related loudness—where all other saliency features are kept constant—even in a complex and distracting environment, explains most of annoyance (Sandrock et al., 2008).

The saliency mechanism could also be evoked to explain observations at a more abstract level of auditory processing. Sounds—or their nomic or symbolic mapping—can also be called salient if incongruent with the context. Such saliency could draw attention to the symbol rather than to the sound features. In the latter case, incongruence of the sound in the scene can enhance detectability (Gygi and Shafiro, 2011). Event-related potential measurements confirm the deviant processing, also with complex sounds, but also show that familiarity with the sound has an effect (Kirmse et al., 2009). A foundation for rapid extraction of meaning from a familiar environmental sound was observed even when sounds were not consciously attended. Outward-oriented mechanisms in turn draw attention to the sound features corresponding to the symbol (see further).

2.3.2 Multisensory Attention

The listener embedded in a real environment—in contrast to experimental conditions—relies on all senses to structure a representation of the environment (Driver and Spence, 1998). One sensory modality could also draw spatial attention to a different modality and even strongly influence the perception itself. This raises the question of whether attention resources are controlled by a supramodal system or by many modality-specific attention systems. In focused attention conditions, judging each signal (sound and vision) separately when incongruent sensory signals occur at the same location is difficult, at least much more difficult than when the incongruent signals come from different spatial locations and attention is divided (Santangelo et al., 2010). A multilevel mechanism of attention with a multimodal component overarching the single sensory component seems the most plausible model given today's knowledge. In the context of assessing the sonic environment, this could be interpreted as a stronger emphasis on visible sources, but at the same time a lower identification probability of a deviant sound experience if this sound comes from the same location as the visual stimulus.

Multisensory attention mechanisms also have a strong temporal component. Sound stimuli presented in temporal congruence with the appearance of a visual target make the visual target pop out of the scene (Talsma et al., 2010). Likewise, visual stimuli that appear—independently of where they appear—at the same moment that a sound could be detected increase the probability that attention will be paid to that sound.

Based on this knowledge on multisensory perception, a long-standing concern of soundscape designers can at least be partly answered: Is it advantageous to hide unwanted sound sources from view? From the attention perspective, one could conclude that provided that the sound is not very salient, and thus is not very likely to attract attention, noticing the sound can be avoided by eliminating visual stimuli that are congruent is space and time with the unwanted sound. Similarly, a wanted sound should be accompanied by a visual stimulus to ensure that it receives proper attention. It should, however, be noted that in case of very salient sounds that will certainly attract attention, the absence of a visual stimulus may come more as a surprise, which may influence appraisal.

2.3.3 Attention to Location

This brings us to the point of binaural hearing. Inhibition of return on location (Mondor et al., 1998) could explain why moving sources or groups of sources of the same kind popping up at different locations might be less easily inhibited by the auditory system and thus continue to attract attention longer than a stationary source. It is known that identity information predominates over location information in auditory memory (Mayr et al., 2011); thus, soundscape appraisal (see the following sections) in itself—in contrast to unmasking—may be less sensitive to aspects of binaural hearing.

Source–listener distance is another aspect of location that might influence attention. As loudness is a primary clue for distance perception, and loudness—or at least loudness change—is known to influence saliency and thus attract attention, there seems to be indirect evidence that sounds from close-by sources would attract more attention. However, we found no experimental evidence that nonfluctuating sounds from a source at a close distance would attract attention more strongly than louder sounds at longer distances.

2.3.4 Voluntary (Endogeneous) Attention and Attention to Memory

Listening in search or story listening involves voluntary (endogeneous) attention focusing grounded in higher-level cognition. It can be shaped by expectations about the place based on prior experience or knowledge, or it can be initially triggered by involuntary attention focusing. Familiarity with the sound is a prerequisite for voluntary attention focusing, yet unfamiliar or incongruent sounds are more likely to attract attention for reasons that can be explained as complex saliency (see above). The interplay between involuntary and voluntary attention results in sustained attention to particular sounds in the sonic environment. Known sounds—which are most likely sounds with strong meaning—could therefore easily be used in soundscape design to draw and maintain attention. Using unfamiliar sounds—as an element of surprise—may need additional context, visual, for example, to ensure that they get attention.

Occasionally, intended activities—and the reason to go to a place—involve listening in search or story listening. For example, one could expect that voluntary attention is focused on natural sounds (birds, breaking waves) if a person is visiting a place to experience nature. However, the complex interplay between expectation, appraisal, and coping may also lead to increased attention focusing on the unexpected or unwanted sound.

The reaction of the brain to sensory stimuli depends on its current state. According to the attention to memory model hypothesis, very similar attention mechanisms are involved in memory tasks, on the one hand, and sensory processing tasks, on the other (Cabeza et al., 2011). Part of the neural circuitry even seems to overlap. This implies additional modulation of overall attention devoted to the sonic environment. Conversely, it also implies that sensory input in general and sound in particular can distract from memory (and cognitive) tasks. Soundscape perception can therefore be different for the same person at different instances, dependent on current activity.

2.4 HOW SOUNDS GET MEANING

The role of audition is not mere information processing but recognition, resulting from bottom-up (signal-driven) and top-down (knowledge-driven) processing. The knowledge-driven component should not be underestimated. The sensory perception could even be regarded as a factor correcting and fine-tuning the mental representation of the (sonic) environment. As such, the meaning of sound(s) could be determined as much by the current state of the mind (emotions included) as it is by the stimulus per se.

2.4.1 Meaning and Associative Memory

Meaning to a large extent depends on the associations a stimulus evokes. The process of attaching meaning to (components of) the sonic environment includes several stages of abstraction. At the lowest level, the association between the sound objects and events they stand for is activated. In case of nomic mapping, the sound and events present consistent information. The event itself produces the sound, for example, a car approaching. In this case, the sound source (e.g., the car) will often be the most important factor in creating meaning. Symbolic mapping relates a sound to an event that does not produce the sound. Symbolic representations allow sharing individual experiential meaning and contribute to the elaboration of social (conventional) meaning. For example, church bells are mapped to an event that is not the ringing of a bell per se, but a socially defined event, such as celebration, in a specific culture.

Meaning extends to more abstract levels by associating the recognized event or source to a larger set of concepts in a somewhat vague way. This vague meaning is sharpened by knowledge of the place (Niessen et al., 2008) and by the most recent meaning attached to the auditory stream. The latter could explain why the path followed by the person experiencing a sonic environment may influence the interpretation and appraisal of a sonic environment.

In order to understand the meaning given to the (components of the) sonic environment, one thus has to understand how associations are learned. An organism learns in order to better predict, prepare for, and anticipate possible futures based on the current situation (Bubic et al., 2010) and to evaluate behavioural options. For the current discourse, a few elements from the multitude of learning theories are extracted. Learning can occur because of prediction error or a teaching signal. From a neurological point of view, there might not be that much difference between both types of learning, but from a sociological point of view, they have very different consequences. For learning from prediction error, the organism must have "lived" the consequence of a wrong prediction of the events occurring in its environment. This makes this type of learning different for different persons, although many of the events that people experience are very similar within a given culture, geographical area, and given generation. However, the influence of culture, geographic area, and generation becomes even more pronounced while learning from peers. Thus, although associative memory is individual, some common features can be expected.

2.4.2 Linguistic Discourse as an Expression of Meaning

Although some scholars may argue that meaning and verbal description are very closely related and can thus be unified, a small distinction remains that could clarify some observations made in soundscape research.

The linguistic label assigned to a sound or the event or source it stands for relies on a complex process, such as categorization and naming, where one category does not depend exclusively on its intrinsic properties, but also on its resemblance to and differences from other categories within the whole classificatory system. Therefore, the same signal can be categorized at different levels of specificity (e.g., traffic, car, sports car; human voices, child voice, my child calling) or along different principles of categorization (source, event such as car breaking or starting, global appraisal, wanted vs. unwanted sound). Thus, not only the context of the observation seems to matter, but also the context in which the meaning of a sound is expressed linguistically.

To complicate matters, the meaning of a word or linguistic expression can also differ between persons. A typical example of relevance in soundscape research is description of tranquility and tranquil area. The French word *calme*, for example, was found to represent different things, depending on the persons asked: a more social interpretation with matching human vocalizations, an evocation of nature and natural sounds, or a notion of quietness and absence of sound (Delaitre, 2013). Thus, a match has to be found between the meaning of words and expressions, on the one hand, and the meaning the sonic environment evokes, on the other hand, in order to understand how persons describe their soundscape experience using a narrative.

2.5 APPRAISAL AND QUALITY JUDGEMENT IN A SOCIOCULTURAL CONTEXT

Appraisal and quality judgement of soundscape form the final step in the analysis. These processes can be regarded from different perspectives. Three different perspectives are discussed below: the perspective of neuroscience reveals how low-level brain functionality could explain why certain sounds are appraised more positively; the perspective of learning and predictability could explain the influence of expectations as well as the preferred level of complexity; and the perspective of coping and behavioural options allows us to view appraisal of the sonic environment in its context in the most holistic way.

2.5.1 Appraisal and Affective Neuroscience

Brain imaging techniques are increasingly used to study how classical psychological concepts are encoded in the brain and to identify connectivity and causal relationships between activation of different brain areas. For the discourse of appraisal and quality judgement, the reward system seems worth looking at. Neuroimaging studies have found that the affective valence of pleasure may be coded in separate networks of brain areas from sensation intensity. The reward system can be described as adding hedonic gloss to the sensation, which could be experienced as conscious pleasure (Berridge and Kringelbach, 2008).

Reward comes in different flavours (Berridge and Kringelbach, 2008): liking, wanting, and learning. Liking is the actual pleasure component of award. At the first level, core liking reactions occur that need not be conscious; at a second level, cognitive brain mechanisms of awareness may elaborate conscious liking from this. Wanting is a motivation for reward that could be a conscious desire to reach cognitive goals, but it also has an incentive saliency component that is not necessarily conscious. Wanting can be, but is not necessarily, linked to liking. Learning includes expectations about future rewards that are learned by association, representation, and prediction. The process can be the result of explicit cognitive reasoning, but it could also rely on implicit knowledge or associative conditioning (Pavlovian associations).

The evolutionary advantage of a reward system is clear when it comes to basic (homeostatic) sensory pleasures such as taste and smell, and in social species the advantage of social pleasure in mate finding and group cohesion is also self-evident. However, it is less clear how higher-order pleasures such as artistic, monetary, altruistic, and so forth, fit in the evolutionary picture. One common view is that these awards depend on learning (the third mechanism). Some of the neural circuitry related to basic pleasure nevertheless seems to overlap.

Research on aesthetic processing of sensory perception can also shed some light on how and why a sonic environment is positively or negatively appraised. Aesthetic processing can be seen as appraisal of valence of perceived objects that comes about through a comparison between subjective awareness of current homeostatic state and exteroceptive perception of objects in the environment (Brown et al., 2011). The basic goal of this comparison circuitry is to identify whether perceived objects will satisfy or oppose our homeostatic needs. For appraisal of the sonic environment only very rarely, it could be expected that the sound object relates directly to a homeostatic need. Therefore, it is worth looking more closely at the social needs already mentioned above. The aesthetic experience of art—including music—may be argued to have social functionality, and therefore, it may have co-opted the basic circuitry used for appraisal in the context of homeostatic fulfilment. Brain imaging experiments indeed show that the same areas of the brain are activated.

Recently, Kuppens et al. (2012) reported highly ecologically valid research into the bidirectional relationship between the way we appraise our (current) environment and how that influences how we feel, plan, and act. Kuppens et al. studied this relationship in the context of core affect, which is defined as an integral blend of the dimensions displeasure–pleasure (valence) and passive–active (arousal) (Russell, 2003). Unlike emotional episodes, which are relatively infrequent, core affect is continually present to self-report.

2.5.2 Appraisal and Predictability

As already mentioned above, prediction is a logical outcome of evolution: because the context-dependent meaning of stimuli changes too frequently, evolution could not rely on instinctive responses and had to turn to associative learning mechanisms to link sensory inputs and behavioural responses. In this evolutionary framework, expected events reward the prediction system with pleasure or aesthetic emotions (Perlovsky, 2006) and a knowledge instinct is developed.

Perception of a visual or sonic environment is thus not a one-way process; the brain is constantly trying to predict the upcoming stimuli. Prediction error causes additional learning and adaptation of the prediction confidence (Winkler et al., 2009). Predictability is appraised as pleasing and aesthetic, yet too much predictability may result in a bored cognitive system. The optimal amount of predictability depends on personal characteristics and mood of the individual. Stimulus complexity and personal experience with this type of stimulus both contribute (Van de Cruys and Wagemans, 2011).

It is worth looking into music research knowledge on predictability and surprise more closely to understand the pleasure in novelty and surprise— or in other words, prediction error—and how it can affect soundscape. In Huron (2006), three kinds of response are identified: prediction, reaction, and appraisal. The prediction response has already been discussed and triggers an aesthetic emotion if the event matches expectations. The reaction response is a fast automatic response that prepares the organism for flight, fight, or freeze in case of surprise, occurring when the event does not match expectations. Finally, the appraisal response is a more leisurely process of consideration and assessment giving positive and negative outcomes. A preference for predictable events (and sounds) is explained by the anticipatory prediction success being misattributed to the stimulus itself. To explain positive emotions associated with surprise, Huron (2006) introduces emotional contrastive valence between the different expectation responses. Events that are welcome or just inoffensive but unexpected can still trigger positive appraisal. But contrastive valence also produces three kinds of pleasurable physiological response: awe, laughter, and frisson. Unexpected events (and sounds) also increase physiological arousal.

An alternative explanation for the inverted U-shape relation of aesthetics with stimulus complexity can be given by merely considering learning (Pearce and Wiggins, 2012). Extremely unpredictable stimuli afford reduced opportunities for learning, while "the learning stimulated by moderate degrees of expectation violation would be pleasurable per se" (Pearce and Wiggins, 2012, page 643).

2.5.3 Appraisal, Coping, and Behavioural Options

From a psychological perspective, appraisal of environmental stressors such as sound is often related to coping opportunities. In a primary appraisal, a person evaluates the situation with respect to its well-being. When the situation is perceived as harmful or threatening, coping resources are assessed in a secondary appraisal step. Although this appraisal and coping theory allows us to model annoyance (Botteldooren and Lercher, 2004; Maris et al., 2007), it seems to be too restricted to negative appraisal to be applicable in its basic form to the soundscape approach, except for restorative soundscapes acting as a coping resource.

The dual-phase appraisal concept can, however, be refined. For this, "coping ability" can be broadened to "opening behavioural options." If something affords behavioural options, it can be regarded as meaningful (Andringa, 2010). Sonic environments that support the behaviour that is instantaneously desired would thus be appraised positively. After a primary appraisal, behaviour options could be assessed and a secondary appraisal might follow. Sounds associated with events that open a desired behaviour option, even if unexpected, are welcomed; sounds related to events that do not prevent the behaviour option could be regarded as inoffensive.

2.6 HOLISTIC MODEL FOR SOUNDSCAPE

Classically, environmental noise has been considered a waste that needs to be prevented or mitigated in volume once the noise-producing activity has been planned. This paradigm could be related to the historical end-of-pipe approach. The soundscape approach introduces a few shifts in this paradigm. Including positive as well as negative environmental sounds in designing high-quality living environments is probably the most easily identified. However, the soundscape approach is also a user-centred approach in line with more general user-centred (product) design. The person-centred view on soundscape therefore allows understanding of many of the ideas and concepts that have been introduced by scholars and practitioners. Finally, the soundscape approach is also an integrated approach including the holistic sensory experience and all different use aspects, such as mobility, recreation, and so forth.

2.6.1 From a Use Case Perspective

In view of the user-centred approach, it is useful to discuss the holistic model for soundscape from a use case perspective. Figure 2.1 shows an example: A person has the intention to move to a place. This creates some expectations that are based on prior knowledge. When the person enters the environment, expectations are fine-tuned by observations. The intention also entails a behaviour that—in this case—is assumed to include listening for particular sounds. Observing these expected sounds leads to a pleasurable experience since it matches expectations. This affects the mood of the person and may lead to new intentions with, for example, a behaviour that

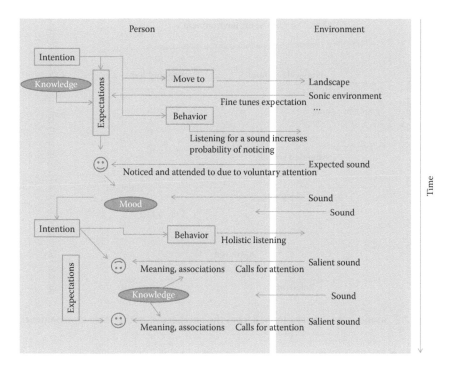

Figure 2.1 An example of a use case of a public space, focusing on environmental sound.

does not include listening in search. A salient sound may nevertheless call for attention. It could be given a meaning that makes the sound disliked. Another salient sound, although it comes as a surprise, may still generate pleasure because the meaning associated with it stimulates the intended behaviour, or at least does not offend it.

This abstract use case can be made more concrete in specific situations, yet it illustrates the complex interplay between different factors and influences.

The sequence of experiences was shown to be rather important since the listener does not respond to the current sensory input only, but also to the current sensory input interpreted and understood within a context that is created by the recent past. At the shortest time frame, meaning (e.g., sound recognition) is given within the framework of very recent experiences (e.g., sounds recently heard). Attention and inhibition of return interplay to avoid focusing on specific sensory inputs, and thus changes and transitions become much more significant than continuous audiovisual stimulation. At a somewhat longer time frame, expectations are fine-tuned and behaviour or even intentions are modified by recent experience. Transitions are therefore caused not only by changes in the environment, but also by changes in the individual and changes in location. This smoothly introduces the important role of accounting for routes and paths travelled.

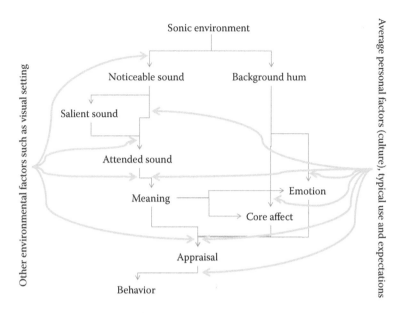

Figure 2.2 Soundscape theory outlined from the perspective of the sonic environment.

2.6.2 From the Perspective of a Sonic Environment

It is also worth looking at the conceptual holistic model for soundscape from a perspective of the sonic environment since designing a soundscape will mainly involve this sonic environment (Figure 2.2). From this perspective, personal traits, preferences, intentions, associations, and so forth, cannot be taken into account, yet the common factor between expected users of this environment can. Based on the overview given above, it can be concluded that foregrounding and backgrounding of specific sounds are important parts of listening. Foregrounding is controlled by attention, yet some sounds will not be noticeable at all, and therefore cannot be attended to. We could call this the undefined background hum. Noticeability of a sound itself depends on personal knowledge and listening capabilities, but from the perspective of the sonic environment, one can only talk about noticeability for the average listener given the environmental context, such as visibility of the sound. Whether noticeable sounds are attended to depends on the saliency of the sound, but also on many personal factors and expectations. For the latter, the typical use of the place, including typical access trajectories followed by the user of the space, is important. Noticeable sounds that do not get noticed blend into the background. Both background(ed) sound and foregrounded sounds contribute to the core affect and emotions evoked by the sonic environment. Foregrounded sounds are expected to have a stronger effect through the meaning and associations they trigger, often related to the source of the sound and the relationship of the average user to that source. Culture—regarded as a common factor in associations—and expectations will play a role. For backgrounded sound the spectrotemporal variations influence core affect and emotions directly through elements of surprise and novelty or continuity. Appraisal—mainly reported appraisal—can be interpreted as a more cognitive evaluation of the sonic environment. Both core affect and meaning influence how the sonic environment is appraised and reappraised in view of behavioural options, for example. Again, the influence of other environmental factors, culture, and expectations is significant.

It should be noted that in the model presented in Figure 2.2, the important feedback paths to modes of listening are not explicated. On an individual basis they are extremely important; however, when looking at soundscape from a perspective of the sonic environment, this iterative detail cannot be taken into account.

2.7 HOW SOUNDSCAPE THEORY CAN AFFECT PRACTICE

In this section the understanding of the processes involved in the creation of soundscape obtained above from reflection on knowledge from psychophysics, neuroscience, psychology, and so forth, is translated to practice. It gives possible directions in measuring and design. Both aspects will be elaborated on in Chapters 5 and 6 and Chapter 8.

2.7.1 Measuring Soundscape

Measuring is about representation (the justification of number assignment) and uniqueness (the representation chosen approaches being the only one possible for the object or phenomenon in question) (*Encyclopædia Britannica*). The main challenge with regard to measuring soundscape is that soundscape is a multifaceted phenomenon and hence cannot be measured with a single number.

2.7.1.1 Measuring People

When measuring people, the investigator wants minimal interference with the test persons. The observation is mainly retrospective unless subtle biomonitoring can be used. This kind of measuring can attempt to capture core affect, appraisal, restoration, and overt behaviour, and thus assesses the soundscape as a whole within a context. This type of measurement fulfils the role of creating a representation perfectly, but it is rather difficult to obtain uniqueness in the measurement. Moreover, this type of measurement should respect the way people are experiencing their environment. So the measurements should characterize not only locations one by one, but also paths between different locations.

The concept of core affect and the associated appraisal of the sonic environment appear in a number of soundscape studies. Depending on the choice of the researchers, the main appraisal dimensions are termed either pleasantness and eventfulness—which match the dimensions of core affect (Axelssön et al., 2010)—or a combination of these dimensions rotated by 45°. Cain et al. (2013) report the dimensions' vibrancy (interpreted as a combination of pleasant and eventful) and calmness (combining pleasant and uneventful). Axelssön et al. (2010) propose to interpret the vibrancy dimension as a continuum from monotonous to exciting, and the calmness dimension spanning from calm to chaotic. One could argue that these dimensions of core affect are related to the person rather than to the sonic environment, but with soundscape interpreted as an object in the mind, this does not pose any problem.

To apply this holistic approach, interviews and questionnaires are the most commonly used tool. To use the results in a planning process, information on the processes discussed above may be gathered: What sounds did people hear (attention process combined with short-term memory [Terroir et al., 2013])? What did these sounds mean to them? How does this relate to expectations concerning the place? This information should be collected after the main appraisal questions in order not to steer the attention process.

2.7.1.2 Measuring with People

Measuring with people implies that the sensory and cognitive capabilities of humans are used to assess the (sonic) environment. The participants are usually in an attentive, analytic listening mode, and thus noticeability and quality of the sound(s) per se are assessed. There is a subtle but rather important difference between this kind of measurement and the measurement of people discussed above: whether or not the sound will actually be noticed in a natural setting with persons engaged in certain activities is no longer considered. To obtain measurements that fulfil the uniqueness criterion, either one has to rely on statistical averaging of the personal factors that might influence the human observation or a master scaling (Lavandier et al., 2012) has to be used to eliminate some of these personal factors by first asking the participants to judge a set of standard stimuli. These stimuli could be either classical pink or white noise samples or reference sonic environments explicitly exhibiting the soundscape features that the research is trying to explore. The latter comes down to calibrating the human as measurement equipment.

The human observer has some capabilities that are hard to mimic using electronics and computational intelligence, for example, the capability to segregate the auditory environment into streams and objects. Thus, questions such as identifying the dominant sound source can easily be answered. Measuring with people, because of the attentive analytic listening mode, is particularly suitable for an analytic description of the soundscape.

An analytic description of the soundscape includes an inventarization of the sounds and the sources producing these sounds. It may also include a description of the quality of the sound, the meaning for that particular group of people, and an indication of congruency. The latter requires the definition of a clear context—sketched in the lab or influenced by the place in field studies—that generates particular expectations.

2.7.1.3 Measuring with Computers in a Human-Mimicking Way

Electronic equipment embodying computational intelligence can mimic the listening capabilities of humans. For the easiest indicators based on level, temporal variability of the level, spectrum, and loudness, each measurement tool produces the same unique outcome, and comparability between sonic environments becomes trivial. However, the representation of the soundscape that is created is rather poor. Information on level, spectrum, and loudness is not sufficient to allow the evaluator or designer to imagine the soundscape.

More advanced, smart sound metres are being developed that allow us to segregate the sound stream into auditory objects (Boes et al., 2012) and label these objects (Boes et al., 2013) taking into account expected sounds at a given location (Krijnders et al., 2010). Besides mimicking the auditory stream segregation, such measurement approaches also could account for the frequency of noticeability or frequency of paying attention to particular sounds (Oldoni et al., 2013). As such, these novel approaches now cover part of the measurements than can be performed with people.

Using electronic equipment has a clear advantage over measuring people: it allows for long-term monitoring. Such monitoring is necessary to study diurnal and seasonal variations in the soundscape. It is also an essential tool to detect novel and unexpected soundscape elements. However, the uniqueness requirement, which is an important factor in measurement theory, is somewhat jeopardized, as less reproducible aspects of human listening are incorporated in measurement equipment.

It should be stressed that research on measuring soundscape either with people or with human mimicking equipment is still ongoing.

2.7.2 Soundscape Design

The goal of soundscape design is to create environmental comfort by influencing the mood, the emotion, the appraisal, and the restoration of persons visiting the place. Based on the soundscape theory explained above, guidelines for future design can be formulated.

2.7.2.1 Designer's Vision and Possible Use of the Place

Modern soundscape design should start from a vision of a place and a soundscape that matches that vision. As urban design is functional design, the use of the space should be accounted for. Running typical use cases (Figure 1.1) should allow the designer to imagine and formulate the expectations of current or future users of the space. It was indeed shown that these expectations may influence the soundscape appraisal or even the mere perception of the sonic environment to a very high degree. As different uses of the space may be envisaged, careful zoning may be needed to match different expectations.

Multimodal aspects—visual, thermal, and so on—should be considered, including the spatial aspect of audiovisual matching. The path usually followed by people visiting the place has to be taken into account since recent experience has a strong influence on listening style, attention paid to sound, recognition, meaning, and appraisal.

This careful initial design phase could lead to formulating requirements for the soundscape:

- Backgrounded: Where the activity and behaviour require the soundscape to remain unattended. One should not notice that any sounds are there.
- Supportive: Where the soundscape enhances the experience and the effect of a visit to a place. The soundscape could improve the mental restoration capacity of the place (see Chapter 3); it could enhance the touristic experience, and so forth, but only as part of a multisensory experience. Specific soundmarks occasionally can attract attention or are expected as part of the experience of a place.
- Focused: Where the soundscape becomes a point of interest in itself. This can be either static, as an acoustic sculpture, or more dynamic, as surprising but pleasing sonic events, where pleasing is defined as supporting or at least not jeopardizing behaviour options.

2.7.2.2 Composition

Once the design goals have been set, the soundscape architect can start to compose the soundscape. In this, the main factors are guiding attention of the visitor and knowing the meaning of the sounds that are integrated in the composition. Attention should be purposely directed to certain components of the sonic environment, to certain sounds, while keeping attention away from unavoidable sounds that are not wanted by the designer. Sounds can be analyzed for their saliency, and for their familiarity, for the most likely users, as both determine the probability that they will be paid attention to. Visual stimuli can create opportunities for wanted sounds to attract attention, but at the same time, they can cause conflict when creating incongruence between visual and auditory stimuli. The probability that an undesired sound will attract attention anyhow, even if its source is not visible, is a crucial factor.

Composing a backgrounded soundscape is conceptually easy and largely boils down to classical noise control. Yet, taking into account saliency of the sound and congruence with the environment could allow reaching a more precise solution.

Supportive soundscapes need not only a careful selection of wanted sounds, but also a careful balance between predictability and novelty. Pleasurable elements of surprise should occur while the user is moving through the space. This allows introducing the desired amount of vibrancy or calmness, matching the whole experience. Composing a supportive soundscape is by far the most challenging.

Composing a soundscape that in itself becomes a point of interest allows for the largest amount of creativity. The auditory experience is dominated by the composition that gets the full attention of the visitor. Thus, the soundscape architect can focus on meaning, emotion, and core affect. Contrastive valence leading to awe, laughter, or frisson can be used. Examples will be given in Chapter 10.

2.8 FUTURE DIRECTIONS AND A WAY FORWARD

Understanding human (auditory) scene analysis and the important role of (auditory) attention in this process allows us to outline better assessment methods and to come to better methodologies for designing desirable soundscapes within a specific context and for a specific use. However, today the knowledge of attention mechanisms and scene analysis applicable to soundscapes has to be inferred from experimental work that uses abstract sounds in a clean context. In this chapter we attempted to do just that. The natural environment is nothing like that: it is governed by complexity, and the biological perception system has evolved to find approximate solutions for achieving goals within this complexity. Experimental research and modelling of attention and scene analysis in this natural environment have to be extended (Lewicki et al., 2014).

It has been pointed out in the previous sections that there are strong individual differences in how a sonic environment is appraised and what meaning is given to the sounds that are noticed within this sonic environment. Cultural elements and age certainly play a role and are mentioned as discriminating factors in appraisal and meaning. Most attempts to understand appraisal of a sonic environment have nevertheless focused on the average person, and indeed some general trends could be discovered from this (e.g., people like natural sounds in a park). Open-ended interviews reveal more details, but still the interviewees are likely to act as local experts assessing the environment in a pseudo-objective and rationalized way, trying to eliminate what they believe is their subjectivity. It may be worthwhile to focus more strongly on the user of a sonic environment and in particular on the diversity in individual traits, beliefs, opinions, and desires. Biomonitoring techniques could be used to assess different responses more objectively than questionnaires, even when it comes to aesthetics or pleasure.

Monitoring and simulating—as a tool for designing—soundscape requires us to account for the way a human listener perceives and understands the sonic environment within a context and use, and with emphasis on personal and cultural differences. This complex process is complicated further by the observation that the use of a place in most cases does not involve attentive listening to the sonic environment. Designing and developing computer software that can mimic this complex process and that can be used in monitoring and modelling is far from easy. Mimicking the human brain in a machine (or even a mouse brain as a starting point) is indeed identified by the European and American research funding agencies as one of the great challenges of this century. Developing machine audition can be seen as part of this challenge. Challenges that could be particularly informative for soundscape include multisensory perception, and in particular multisensory attention mechanisms; introducing learned context awareness; and modelling a biological plausible reward system, including serotonin and dopamine effects (Weng, 2013). The latter could lead to adding appraisal to machine audition in ways that are not foreseeable today.

REFERENCES

Andringa, T. C. (2010). Audition: From sound to sounds. In *Machine Audition: Principles, Algorithms and Systems*, ed. W. Wang, 80–105. IGI Global Press, 532 pages, ISBN-13: 9781615209194, August 2010.

Axelssön, O., Nilsson, M. E., and Berglund, B. (2010). A principal components model of soundscape perception. *Journal of the Acoustical Society of America*, 128, 2836–2846.

Berridge, K. C., and Kringelbach, M. L. (2008). Affective neuroscience of pleasure: Reward in humans and animals. *Psychopharmacology*, 199, 457–480.

Boes, M., Oldoni, D., De Coensel, B., and Botteldooren, D. (2013). Attention-driven auditory stream segregation using a SOM coupled with an excitatory-inhibitory ANN. In *2012 International Joint Conference on Neural Networks (IJCNN 2012)*, Brisbane, Queensland, Australia, 2012, p. 8.

Boes, M., Oldoni, D., De Coensel, B., Botteldooren, D. (2012). A biologically inspired recurrent neural network for sound source recognition incorporating auditory attention. Presented at Proceedings of IJCNN, Dallas, TX, August 4–9 2013.

Botteldooren, D., and Lercher, P. (2004). Soft-computing base analyses of the relationship between annoyance and coping with noise and odor. *Journal of the Acoustical Society of America*, 115(6), 2974–2985.

Bregman, A. S. (1994). *Auditory Scene Analysis: The Perceptual Organization of Sound*. Cambridge, MA: MIT Press.

Brown, S., Xiaoqing, G., Tisdelle, L., Eickhoff, S. B., and Liotti, M. (2011). Naturalizing aesthetics: Brain areas for aesthetic appraisal across sensory modalities. *Neuroimage*, 58(1), 250–258.

Bubic, A., von Cramon, D. I., and Schubotz, R.I. (2010). Prediction, cognition and the brain. *Frontiers in Human Neuroscience*, 4, 25.

Cabeza, R., Mazuz, Y. S., Stokes, J., Kragel, J. E., Woldorff, M. G., Ciaramelli, E., Olson, I. R., and Moscovitch, M. (2011). Overlapping parietal activity in memory and perception: Evidence for the attention to memory model. *Journal of Cognitive Neuroscience*, 23(11), 3209–3217.

Cain, R., Jennings, P., and Poxon J. (2013). The development and application of the emotional dimensions of a soundscape. *Applied Acoustics*, 74, 232–239.

Cusack, R., Decks, J., Aikman, G., and Carlyon, R. P. (2004). Effects of location, frequency region, and time course of selective attention on auditory scene analysis. *Journal of Experimental Psychology: Human Perception and Performance*, 30, 643–656.

Delaitre, P. (2013). Caracterisation des zones calmes en milieu urbain: Qu'entendez-vous par zone calme? PhD thesis, Université de Cergy-Pontoise, France.

Driver, J., and Spence, C. (1998). Attention and the crossmodal construction of space. *Trends in Cognitive Sciences*, 2(7), 254–262.

Escera, C., Alho, K., Winkler, I., and Nätänen, R. (1998). Neural mechanisms of involuntary attention to acoustic novelty and change. *Journal of Cognitive Neuroscience*, 10, 590–604.

Fritz, J. B., Elhilali, M., David, S. V., and Shamma, S. A. (2007). Auditory attention: Focusing the searchlight on sound. *Curr. Opin. Neurobiol.*, 17, 437–455.

Gaver, W. W. (1993a). What in the world do we hear? An ecological approach to auditory event perception. *Ecological Psychology*, 5(1), 1–29.

Gaver, W. W. (1993b). How do we hear in the world? Explorations in ecological acoustics. *Ecological Psychology*, 5(4), 285–313.

Genuit, K., and Fiebig, A. (2006). Psychoacoustics and its benefit for the soundscape approach. *Acta Acustica united with Acustica*, 92(6), 952–958.

Gygi, B., and Shafiro, V. (2011). The incongruency advantage for environmental sounds presented in natural auditory scenes. *Journal of Experimental Psychology—Human Perception and Performance*, 37, 551–565.

Hellman, R. P. (1982). Loudness, annoyance, and noisiness produced by single-tone-noise complexes. *The Journal of the Acoustical Society of America*, 72(1), 62–73.

Huron, D. (2006). *Sweet Anticipation: Music and the Psychology of Expectation*. Cambridge, MA: MIT Press.

ISO 1996-1. (2003). Acoustics—Description, measurement and assessment of environmental noise—Part 1: Basic quantities and assessment procedures.

Kaplan, R., and Kaplan, S. (1989). *The Experience of Nature: A Psychological Perspective*. Cambridge: Cambridge University Press.

Kayser, C., Petkov, C., Lippert, M., and Logothetis, N. K. (2005). Mechanisms for allocating auditory attention: An auditory saliency map. *Current Biology*, 15, 1943–1947.

Kirmse, U., Jacobsen, T., and Schröger, E. (2009). Familiarity affects environmental sound processing outside the focus of attention: An event-related potential study. *Clinical Neurophysiology*, 120(5), 887–896.

Krijnders, J. D., Niessen, M. E., and Andringa, T. C. (2010). Sound event recognition through expectancy-based evaluation of signal-driven hypotheses. *Pattern Recognition Letters*, 31, 1552–1559.

Kuppens, P., Champagne, D., and Tuerlinckx, F. (2012). The dynamic interplay between appraisal and core affect in daily life. *Frontiers in Psychology*, 3, 1–8.

Lavandier, C., Barbot, B., Terroir, J., and Schuette, M. (2012). Calibration of subjects with master scaling: An application to the perceived activity disturbance due to aircraft noise. *Applied Acoustics*, 73, 66–71.

Lavie, N. (2010). Attention, distraction, and cognitive control under load. *Current Directions in Psychological Science*, 19, 143–148.

Lavie, N., and Tsal, Y. (1994). Perceptual load as a major determinant of the locus of selection in visual attention. *Perception and Psychophysics*, 56, 183–197.

Lewicki, M. S., Olshausen, B. A., Surlykke, A., and Moss, C. F. (2014). Scene analysis in the natural environment. *Frontiers in Psychology*, 5, 199. doi: 10.3389/fpsyg.2014.00199.

Maris, E., Stallen, P. J., Vermunt, R., and Steensma, H. (2007). Noise within the social context: Annoyance reduction through fair procedures. *Journal of the Acoustical Society of America*, 121, 2000–2010.

Mayr, S., Buchner, A., Moller, M., and Hauke, R. (2011). Spatial and identity negative priming in audition: Evidence of feature binding in auditory spatial memory. *Attention Perception and Psychophysics*, 73, 1710–1732.

McAdams, S. (1993). Recognition of sound sources and events. In *Thinking in Sound: The Cognitive Psychology of Human Audition*, ed. S. McAdams and E. Bigand. Oxford: Clarendon Press.

Mondor, T. A., Breau, L. M., and Milliken, B. (1998). Inhibitory processes in auditory selective attention: Evidence of location-based and frequency-based inhibition of return. *Perception and Psychophysics*, 60, 296–302.

Niessen, M. E., van Maanen, L., and Andringa T. C. (2008). Disambiguating sounds through context. In *2008 Second IEEE International Conference on Semantic Computing (ICSC)*, 2008, pp. 88–95.

Oldoni, D., De Coensel, B., Boes, M., Rademaker, M., De Baets, B., Van Renterghem, T., and Botteldooren, D. (2013). A computational model of auditory attention for use in soundscape research. *Journal of the Acoustical Society of America*, 134, 852–861.

Payne, S. R. (2013). The production of a perceived restorativeness soundscape scale. *Applied Acoustics*, 74, 255–263.

Pearce, M.T., and Wiggins, G.A. (2012). Auditory expectation: The information dynamics of music perception and cognition. *Topics in Cognitive Science*, 4, 625–652.

Perlovsky, L. I. (2006). Toward physics of the mind: Concepts, emotions, consciousness, and symbols. *Physics of Life Reviews*, 3, 23–55.

Prime, D. J., Tata, M. S., and Ward, L. M. (2003). Event-related potential evidence for attentional inhibition of return in audition. *Neuroreport*, 14, 393–397.

Russell, J. (2003). Core affect and the psychological construction of emotion. *Psychological Review*, 110(1), 145–172.

Sandrock, S., Griefahn, B., Kaczmarek, T., Hafke, H., Preis, A., and Gjestland, T. (2008). Experimental studies on annoyance caused by noises from trams and buses. *Journal of Sound and Vibration*, 313(3), 908–919.

Santangelo, V., Fagioli, S., and Macaluso, E. (2010). The costs of monitoring simultaneously two sensory modalities decrease when dividing attention in space. *Neuroimage*, 49(3), 2717–2727.

Shamma, S. A., Elhilali, M., and Micheyl, C. (2011). Temporal coherence and attention in auditory scene analysis. *Trends in Neuroscience*, 34, 114–123.

Snyder, J. S., and Alain, C. (2007). Sequential auditory scene analysis is preserved in normal aging adults. *Cerebral Cortex*, 17(3), 501–512. doi: 10.1093/cercor/bhj175.

Talsma, D., Senkowski, D., Soto-Faraco, S., Woldorff, M. G. (2010). The multifaceted interplay between attention and multisensory integration. *Trends in Cognitive Sciences*, 14(9), 400–410. doi: 10.1016/j.tics.2010.06.008.

Terroir, J., De Coensel, B., Botteldooren, D., and Lavandier, C. (2013). Activity interference caused by traffic noise: Experimental determination and modeling of the number of noticed sound events. *Acta Acustica united with Acustica*, 99(3), 389–398.

Truax, B. (2001). *Acoustic Communication*. 2nd ed. Westport, CT: Ablex.

Van de Cruys, S., and Wagemans, J. (2011). Putting reward in art: A tentative prediction error account of visual art. *Iperception*, 2, 1035–1062.

Weng, J. (2013). How the brain-mind works: A two-page introduction to a theory. *Brain-Mind Magazine*, 2(2).

Winkler, I., Denham, S. L., and Nelken, I. (2009). Modeling the auditory scene: Predictive regularity representations and perceptual objects. *Trends in Cognitive Sciences*, 13, 532–540.

Chapter 3

Soundscapes, Human Restoration, and Quality of Life

Irene van Kamp,[1] Ronny Klæboe,[2] A. Lex Brown,[3] and Peter Lercher[4]

[1]National Institute for Public Health and the Environment, Bilthoven, The Netherlands

[2]Institute of Transport Economics, Oslo, Norway

[3]Urban Research Program, Griffith School of Environment, Nathan, Griffith University, Brisbane, Australia

[4]Department of Hygiene, Microbiology and Social Medicine, Division of Social Medicine, Medical University of Innsbruck (MUI), Innsbruck, Austria

CONTENTS

3.1 INTRODUCTION

There is increasing interest in human perception and enjoyment of the acoustic environment experienced outdoors in urban, rural, and natural or wilderness areas. In the past decade, work in the field of *soundscapes* has been on understanding the relationship between people and their acoustic environment, examining the sounds that people value and their reaction to them within specific contexts of place and activity. This work has also been concerned with the potential and utility for management, conservation, and design of the acoustic environment to increase human enjoyment. Research over the past 30 years or so has documented the long-term health effects of noise for particular health outcomes. Now there is increasing interest in the idea that the acoustic environment may affect human well-being, not only at the high levels of exposure considered in noise management, and not only where the sounds result in human discomfort and adverse health effects, but also where the acoustic environment is pleasant and of high quality. In this chapter we explore the notion that access to high-quality acoustic environments may have a role in well-being, quality of life, and environmental health—*intrinsically,* or by way of *mediation, moderation, or buffering*—through some restorative and health- and well-being-promoting mechanisms. The limited literature available on soundscapes and restoration, and possible models for their interconnection, is discussed. This review necessarily intersects with other literature that examines links between, on the one hand, landscape, green space, open space, countryside, and recreation, and on the other, well-being, quality of life, and environmental health. *Monetarization* of health and well-being effects of high acoustic quality will also be touched upon. We suggest that further work is needed, even at the conceptual level, to increase our understanding of relationships between soundscape and health.

3.2 ACOUSTIC ENVIRONMENT AND HEALTH

In Chapter 1, it was noted that the acoustic environment of outdoor space has, to date, largely been considered the province of those charged with environmental noise management. Noise management addresses the sounds of discomfort and, nearly always, sounds of higher intensities, such as those from air and surface transport sources. It focuses on managing this sound as a waste product (see Table 1.1) and as a hazard to health. We begin this chapter by a brief review of the health effects of environmental noise, and a model of potential mechanisms by which high levels of sound in the acoustic environment may affect health. Then we extend this somewhat restricted perspective of the relationship between the acoustic environment and human health. The reconceptualizing of environmental sound as a resource, rather than as waste only, promotes consideration of the acoustic environment influencing human health and well-being, not just when levels of sound are intense and cause human discomfort, but also where the acoustic environment is of high quality. Here we explore the notion that access to high-quality acoustic environments may have a role in people's well-being, quality of life, and environmental health. This may be through the intrinsic experience provided by high-quality soundscapes in outdoor spaces, or it may be by way of mediation for people otherwise exposed to adverse environments. The pathway for such influence on human health and well-being would be through some *restorative* mechanisms, or health and well-being-promoting mechanisms.

We canvas available literature on soundscapes and restoration and possible models for their interconnection. This literature is limited, and we focus on conceptual issues and potential mechanisms—illustrated by data when available. The limited literature on the acoustic environment and restoration has encouraged us to examine appropriate models from analogous literature, such as that on green space and health.

3.2.1 Health Effects of Environmental Noise

Noise effects research in the past 40 years has been focused primarily on the adverse health effects of unwanted sound, or noise. Research in this field has been extensive. Chronic exposure to environmental noise has been linked to a range of well-documented health effects. Figure 3.1 shows the potential mechanisms by which noise can lead to health problems. The model is based on a publication of the Health Council of the Netherlands (HCN) [1] and is one of

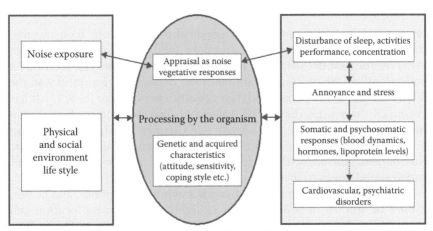

Dynamic demographic, social, cultural, technological and economic environment.

Figure 3.1 Conceptual framework for effects of high levels of sound on health. (From Health Council of the Netherlands (HCN), Public health impact of large airports, HCN Report 1999/14E, The Hague, September 2, 1999, http://www.gr.nl/pdf.php?ID = 19&p = 1.)

the prevailing approaches to noise and health built on a cognitive emotional stimulus–response model. In the stimulus–response approach the emphasis is on threshold levels, norms, and interventions aimed at reducing levels. The model assumes that most effects are a consequence of the appraisal of sound as noise. It is generally assumed that stress responses play an important role in the process by which environmental noise leads to health effects. However, sound can increase a person's alert response and activity level due to interactions of the acoustic nerve with other parts of the central nervous system, and this is particularly relevant during sleep. Noise exposure is associated with annoyance, sleep disturbance, and activity disturbance and stress responses. These underlie instantaneous effects of high levels of sound that include blood pressure increases and increased secretion of cortisol—responses that are considered risk factors for cardiovascular diseases and mental pathology. Responses are partly dependent on the noise characteristics of frequency, intensity, duration, and meaning, and partly on nonacoustical aspects, such as context, attitude, expectations, fear, noise sensitivity, and coping strategies. Van Kamp et al. [2] provide a brief review of current knowledge in this field.

These health effects result from higher levels of exposure to environmental noise, such as that experienced by people living near, inter alia, surface and air transport sources. More than half of the world's population already lives in urbanized areas, and in a further urbanizing world, the environmental noise is increasing given the continuing growth in both surface and air transport modes [3].

3.2.2 Soundscape and Potential Beneficial Health Effects

Interest in human perception and enjoyment of the outdoor acoustic environments in urban, rural, and natural or wilderness areas are growing [4–7]. In the past decade, work in the field of soundscapes has focused on understanding the meaning of sound in the outdoor environment and the relationship between people and their acoustic environment [8–9, 11]. It has examined the sounds that people value and their reaction to them within specific contexts of place and activity (see Chapter 1 and [10]). It has also been concerned with management, conservation, and design of the acoustic environment, including the potential for both protection and creation of varied soundscapes to increase human enjoyment [12–17].

The relationship between soundscapes and human response, and health, cannot be studied in isolation, as it is highly dependent on location and functions, on human activities and behaviour, and on context. This is illustrated in Figure 3.2, where the experience of soundscapes leads to direct and indirect human response outcomes on quality of life and health.

The association between soundscape and health suggested in this figure may not be immediately obvious, but it is useful to explore the potential of such a linkage, including models of feasible mechanisms. Key concepts are

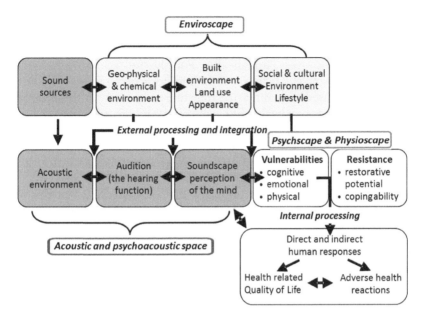

Figure 3.2 A block diagram of the soundscape perception process, moderated by context leading to direct and indirect human responses. (From Lercher, P., and Schulte-Fortkamp, B., Soundscape of European cities and landscapes: Harmonising, in *Soundscape of European Cities and Landscapes*, ed. J. Kang et al., COST, http://noiseabatementsociety.com/wp-content/uploads/2014/01/COST%20 eBOOK%20DOCUMENT%20tryout%2012_5_13%20Final-REVISE19.pdf.)

environmental and acoustic quality, quality of life, and restoration. The important question is: Does the (acoustic) environment contribute (and, if so, how) to our well-being and health by enabling us to restore from (daily) stress?

There is increasing attention to the restorative function of quiet areas on (mental) health, but only a few studies place the relationship between noise and (mental) health in a broader context of soundscapes and environmental quality [19–21]. The available evidence is based on laboratory studies and a handful of epidemiological studies [22]. These address the restorative effects of natural recreational areas in the first place outside the urban environment [23–25]. As a consequence, we do not have evidence as yet as to prerequisite characteristics of urban environments in order for them to contribute to restoration after stress.

3.2.3 Method

A systematic literature search[1] was performed for the specific association between soundscapes and health and well-being. Scopus databases were

---

[1] To improve readability, all the available review material is not included in this chapter. An extensive list can be requested from the first author.

searched to identify relevant peer-reviewed English, German, and Dutch studies published in the period between April 2000 and April 2013. A wide range of keywords was used related to tranquillity, quiet, public or green space, restoration, stress, and health. For the full search strategy, see Annex 1. Several recent reviews, including both peer-reviewed and grey literature, were used to summarize the state of the art on the beneficial effects of green space on health and well-being, and the role of physical and social aspects in the association between sound/noise and well-being.

For economic evaluation of soundscapes, a separate review was carried out covering a period between 2010 and 2014. The full search strategy is in Annex 2.

The literature on the beneficial effects of sound is scarce. For this reason, we consider it appropriate to also examine the much broader, and rapidly growing, literature on the health-promoting effects of green space, as literature from this analogous area can, at least, be used as a point of departure for building knowledge of the relationships between soundscapes and health. The green space, recreation, and environmental health literature currently has models, assumed mechanisms, conceptual approaches, and data on relationships between being in, or undertaking activities in, such places and people's health and well-being. Study of the association between soundscapes, restoration, and well-being can only profit from building on this literature, and also by contextualizing soundscape within a broader physical and social environmental experience.

3.3 CONCEPTUAL MODELS AND MECHANISMS

An integrated approach is needed to further the field of soundscapes and health. In his groundbreaking publication, Kaplan [26] pointed out, and warned, that integrated models or approaches must be seen as metaphors that present complex connective relationships in a simplified manner, and that they should be seen as cartoons that describe the essence of a theory. Their successful assessment and application, however, is dependent on the availability of data and statistical analytical methods. Models currently available in the literature offer, from different perspectives, various components necessary for the study of social, physical, and spatial aspects of the impact of environment on health [27]. However, integration between them is far from a reality. With this in mind, potentially relevant models and approaches are presented below, not in any particular order, and not differentiating between their fields of origin. Where possible, they are presented in conjunction with exemplary evidence. It is suggested that these models may eventually prove to be the building blocks from which an integrated model of the relationships between soundscapes and human health and well-being can be developed.

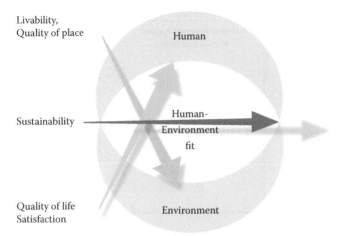

Figure 3.3 Person–environment fit in its time frame. (From van Kamp, I., et al., *Landsc. Urban Plan.*, 65(1–2): 5–8, 2003.)

3.3.1 Interaction of Person and Environment

Here, person–environment interaction is taken as the starting point—based on the model by Herranz-Pascual et al. [10] in Chapter 1. The relationship between environmental quality, quality of life, and restoration is not static, but a transactional process in which causality is harder to define. Which comes first? Disruptions or changes in the environment, as well as in human behaviour, can influence the human–environment fit, as depicted in Figure 3.3.

The outcomes of this human–environment fit can be described in terms of health and well-being, behaviour, stress, and quality of life, and restoration can be a part of this process.

3.3.2 Integrated Model on Nature and Health

Several models from the broad literature on the health benefits of green areas were recently summarized by Murel et al. [29] and Lachowycz and Jones [30]. Already in 2004, the HCN [31] integrated available models in one framework, which is still used as the point of departure for studies into the relationship between natural environments and health and well-being.

The model in Figure 3.4 assumes that exposure to a natural environment leads to health and well-being via a process of restoration from stress and attentional fatigue, from undertaking physical activity, and from social contacts, as affected by life-stage issues. Others include an additional factor in the model: contribution through a reduction in environmental exposures, for example, noise, air pollution, heat, or light (UV). Although it is

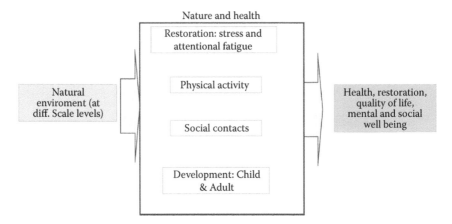

Figure 3.4 HCN source: conceptual framework for nature and health. (From Health Council of the Netherlands and Council for Spatial Environment and Nature Research, *Natuur en gezondheid, Invloed van natuur op sociaal, psychisch en lichamelijk welbevinden*, HCN Report 2004/09, A02a, Den Haag: Gezondheidsraad en RMNO, 2004.)

known that access to green areas in urban contexts can mitigate the effects of such environmental stressors and pollutants, this additional factor in the model can also work the other way around. For example, a natural area may lose its restorative potential by the dominance of some unwanted noise source, such as snowmobiles [32] or aircraft [7, 33].

3.3.3 Model Pertaining to a Need for Quietness

An example of an approach describing behaviour as an outcome is the model (Figure 3.5) of Booi and van den Berg [12]. They summarized three

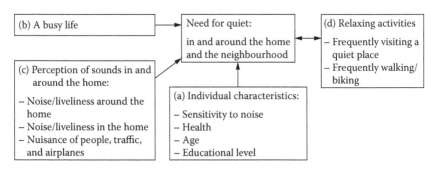

Figure 3.5 A model on the need and whether people lead a busy life or not (b). All of these are assumed to affect the need for quiet and together predict the engagement (d) in relaxing activities for quiet. (From Hester, B., and van den Berg, F., *Int. J. Environ. Res. Public Health*, 9: 1030–1050, 2012.)

categories of variables, each of which might influence the need for quietness: those pertaining to individual characteristics (a), a busy life (b), and the perception of sounds around the home and in the immediate residential area (c). These aspects together predict the engagement in relaxing activities (d).

3.3.4 Mechanism of Stress and Restoration

The term *restoration* was first introduced by Kaplan and Kaplan [34] in their attention restoration theory (ART). Research into the mechanism confirmed that people can concentrate better after spending time in nature or looking at scenes of nature [34, 35]. A dictionary definition of the term *restoration* refers to the act of bringing something back to a former place, state, station, or condition; the fact of being restored, renewed, or reestablished. Hartig and coworkers [22, 36] and Korpela coworkers [37, 38] define a restorative experience as the degree in which an environment can aid recovery from mental fatigue and attentional capacities.

Staats [39] and Wallenius [40] refer to inhibited restoration or a lack of stress recovery and consider restoration important for mental as well a physiological recovery and long-term health. In relation to noise, these authors showed that perceived control and noise sensitivity were important moderators. Gidlöf-Gunnarsson and Öhrström [41, 42] define psychological restoration as having access to places where one can relax, unwind, and feel content and undisturbed.

Attention restoration theory [23, 35, 43, 44] distinguishes four necessary components of restoration:

1. Being away: Psychological distance from the demands and routines in which people use the directed attention capacity.
2. Fascination: Attention is captured by aspects of flora and fauna.
3. Extent: Degree and scope of exploration on the environment.
4. Compatibility: Match between what the person wants to do and must do.

The concept of restoration could potentially link with health through the restorative function of people's experience of areas of high acoustic quality, particularly through its emphasis on context and meaning. There is some indication that certain places contribute to restoration. Often, it is assumed that this is related to aspects of quiet and green, but evidence is still limited and insight into the mechanisms needs further attention. Most studies in this domain [45–51] address the restorative effects of natural recreational areas outside the urban environment [41].

A question is whether natural areas within, and in the vicinity of, urban areas similarly contribute to psychophysiological and mental restoration after stress [52]. Also, does restoration require the absence of urban noise? In an early study of Ulrich et al. [53] that included physiologic measures, it was found that exposure to pure nature sites resulted in a slightly better restoration than exposure to urban restoration sites, and that high levels of traffic noise were associated with poor recovery. The differences were small, but the trends were clear. Besides potential immediate restorative effects, there may be long-term effects of access to environmental amenities in the immediate living environment. Dutch cross-sectional studies [22, 45, 47] found that residents in green neighbourhoods report a better general health. Others have described the buffering effect between stress and health of the availability of green [48, 54]. Do natural environments (micro and macro) positively influence long-term general health and well-being, and which environmental aspects are important? The soundscape approach is aimed at location-specific acoustical quality and its immediate effects (see Chapter 2 for some qualification on this) on people's perception and well-being. The long-term health effects of access to areas of high acoustic quality clearly need to be studied in more depth.

3.4 RESTORATION ASSOCIATED WITH SOUNDSCAPES

The available evidence on restoration associated with the acoustic environment, and literature on green areas, suggests that we should consider at least two different mechanisms:

- Type 1 restoration, in which a high-quality acoustic environment intrinsically provides restoration *by an immediate pathway*—in the same way as green or natural areas, wilderness, or urban environments with natural elements are assumed to do. Figure 3.6 shows examples of environments of high acoustic quality where one can postulate that type 1 restorative effects could occur through spending time in such places [45].
- Type 2 restoration refers to the effect of availability (knowledge) of a high (better)-quality acoustic environment to a person who otherwise is subject to adverse effects of noise, usually in their home environment. Data are presented below where individuals had experienced adverse levels of noise in their dwellings, but then reported lower adverse effects if their dwelling also has a "quiet side"; that is, their dwelling also provided access to a more positive acoustic environment. Individuals have also reported lower adverse effects of noise in their dwellings if they have access to, or even knowledge about, nearby green areas. Type 2 restoration can be considered restoration

Figure 3.6 Environments with high acoustic quality. (Photo's 2-5 by A. L. Brown.)

by way of mediation or by way of moderation, also referred to as a *buffer effect* between stress and health, as described by Wells et al. [53, 54]. Available evidence for type 2 restoration is detailed below.

3.4.1 Effect of a Quiet Side of the Dwelling

A recent review in Dutch [55] summarized the health effects of availability of a quiet side of a dwelling. Several studies [56–62] show that people who have access to a quiet side in their dwelling report less noise annoyance than their counterparts who have no quiet side available, and a beneficial effect on the quality of their sleep. Results of the few and often small intervention studies showed that the sound quality both indoors and outdoors was experienced as better after a quiet side had been created, but the long-term health effects of this have not been studied. This is in line with earlier work addressing the (health) effects of a number of residential features, which indicated that having a sleeping room on the quiet side of the dwelling has a beneficial effect on annoyance, sleep, and blood pressure [58]. The association with blood pressure might be mediated by sleep quality. Temporary respite from environmental noise could reduce the negative effects on health and well-being [59]. We have to bear in mind that most studies of these effects have been cross-sectional, which makes it hard to draw conclusions about the direction of the association. The assumed process of restoration [23] can only be studied in an experimental or longitudinal study design.

The definition of a quiet side has varied between studies. It is often expressed as the difference in levels between the highest and lowest exposed façades (in $L_{Aeq24hr}$ or L_{den}). In the EU guidelines (END), a quiet façade is defined as having a relatively low noise exposure—at least 20 dB below the most exposed façade [63]. In other documents, a quiet side is defined as a total exposure level, expressed in $L_{Aeq24hr}$ below a certain level (e.g., 48 dB) (see, e.g., [61]).

3.4.2 Access to Quiet Places Near the Dwelling

The Health Council of the Netherlands (HCN) [46] stresses that quiet and peace could enhance health and well-being by people spending time in a garden or other space where they can relax. Results of a Swedish study have shown that access to green space, as well as the quality of this green space, in the immediate residential area, moderated the relation between a quiet side and annoyance [41, 42]. These results were in line with findings of Klæboe et al. [64–66], who concluded that the acoustic quality in the vicinity (within 75 m) of the dwelling, operationalized in the degree in which a neighbourhood soundscape was more or less noisy than what is usually associated with the residential noise level, moderated the association between road traffic noise and annoyance. The presence of quiet places did not affect the annoyance. Italian research [67] confirmed that the perception of the quality of gardens positively influenced the association between noise and annoyance. Li et al. [68] showed that having a view to green space or to "blue" moderated the association between road traffic noise and annoyance in Hong Kong.

3.4.3 Access to Green and Tranquil Space in the Wider Area

Results of a study in Amsterdam showed that the need for quiet was strongly related to noise sensitivity and the perception of sound [12]. People who perceived the sound environment as a negative factor, such as noise from transportation and people, reported a higher need for quietness, but when the sound environment was labelled as positive (such as perceived liveliness at home or in the neighbourhood), it seemed to reduce that need. The need for quiet was also associated with educational level: the higher the education, the greater the need for quietness. These findings match those reported by van Kempen et al. [69]. Analysis of data from a neighbourhood ($N = 3600$) survey in the Netherlands showed that the score on a mental health index (Rand 36) was associated with a larger need for quiet, more visits to quiet areas, and lower satisfaction with access to quiet.

Analysis on data sets available from the Schiphol longitudinal study before and after the opening of a fifth runway [70] showed that a high score on the negative dimension of the perceived soundscape index [71] was associated with severe annoyance from all noise sources, a lack of vitality, psychological distress, the prevalence and incidence of depression and anxiety, a higher need for quiet, less visits to quiet, and dissatisfaction with access to quiet (after adjustment for age, gender, education, ethnicity, and air traffic noise exposure level). A positive perception was associated with significantly less annoyance from all sources, a higher level of vitality, less need for quiet, and more visits to quiet, and a lower dissatisfaction with access to quiet.

A more integrated and contextual approach would be needed to unravel the complex and dynamic process of the interaction between people and their environments. People with varying susceptibilities need to be compared. When studying the mental effects of noise in susceptible groups, two things have to be taken into account: the broader context in terms of available amenities and differential benefits of these amenities in terms of restoration [72–74].

Very few studies have evaluated the physical and spatial characteristics of areas with good sound quality [75]. As part of the Quiet Places Project in Amsterdam, people were asked to name a quiet place in their neighbourhood or elsewhere [12]. Most often mentioned was a courtyard with a chapel that is located just around the corner of the busiest shopping street of Amsterdam. In response to the question "What characterizes your favourite, nearby quiet area?" the presence of green and water was mentioned most often (96% of the respondents), followed by cleanliness/well kept (77%), nice colours (74%), spacious (68%), and nice odour (52%). Suburban residents characterized a quiet area differently than city dwellers. Earlier, De Coensel and Botteldooren [76] developed a set of criteria for the quality of quiet rural areas. According to them, these spaces may be characterized by a combination of acoustical criteria, such as relatively low sound levels and the relative absence of nonfitting sounds, and nonacoustical criteria, such as the presence of natural elements within the visual scene.

Pheasant and colleagues [77–78] studied the characteristics of tranquil places ("space that can facilitate a state of tranquillity, suggesting reconciliation between mental space and the physical and social spaces in which we live and work"). In order to assess people's perceived tranquillity of a location, they presented visual and acoustic data captured from 11 rural and urban landscapes to volunteers. Maximum sound pressure level (L_{Amax}) and the percentage of natural features of a location showed to be key factors influencing perceived tranquillity. In a later phase of the study, a range of man-made features that directly contributed to the overall visual context of the environment appeared to be of importance. Features under study were listed buildings, religious and historic buildings, landmarks, monuments, and man-made elements of the landscape.

Since indications exist that spending time in areas with good sound quality is beneficial for health and well-being by offering psychophysiological recovery (restoration) from stressors such as noise and mental fatigue, we could also learn from the results from studies that investigate (features in the) built environments that can promote restoration. Lindal and Hartig [79], for example, generated 145 images of streetscapes. Participants had to rate these images with regard to the likelihood of restoration, fascination, being away, and preference. The results showed that attributes such as building height and façade details influenced people's judgements regarding restoration likelihood for urban residential streetscapes (see also [36] and [44]).

3.5 ECONOMIC PERSPECTIVE ON SOUNDSCAPES AND RESTORATION

Accessible higher-quality indoor and outdoor soundscapes can be regarded as local public goods in the sense that their use is nonexclusive and, up to some number of users, nonrivalled [80]. High-quality soundscapes bestow benefits on current and potential users [42]. Economists put emphasis on relative scarcity and substitutability of goods. Public goods, including soundscapes, have economic value, potentially measurable in money terms, even if we do not pay directly for their use. However, the money value of a good is not absolute, but relative to its scarcity or abundance and the ease with which it can be replaced with an alternative.

The importance of, for example, a quiet area with potential to provide restoration will thus depend on how many other parks there are in the vicinity and whether the same need can be served or fulfilled by quiet rooms, travelling to more distant recreational areas, or meditation in a suitable indoor environment. The loss of a quiet park is more problematic if no substitutes exist. Questionnaires and surveys that attempt to elicit the value of soundscape thus need to extract information on alternatives and other choices available.

The traditional noise control approach focuses on areas exceeding certain noise levels using regulation (noise zones, limits, and guidelines) and financial disincentives (polluter pays) to limit adverse health effects. Promoters of urban greenery, and positive soundscapes, focus instead on the value of positive urban environments in attracting people, businesses, and economic activity. When some cities are successful in improving the quality of their urban environment, attracting businesses and a high-quality workforce, market forces induce the neighbouring cities to catch up if they wish to remain competitive [81].

Many authorities make use of cost–benefit analyses to ascertain whether measures, like those improving the urban soundscapes, are worth more than they cost [82, 83]. For noise control measures, economic values of noise reductions are available obtained from standard unit prices, for example, for a given per decibel reduction, multiplied by the number of affected persons or dwellings. However, when noise control measures have nonacoustic effects, these should also be assessed in economic terms [84]. To assess economic benefits provided by areas suitable for restoration, all qualities (acoustic, aesthetic, and others) should be calculated in money terms, for a complete (social) cost–benefit analysis.

The hedonic pricing method is often used to assess the monetary value of local public goods, like noise/soundscape quality [85–88]. In the hedonic pricing approach, the price differential between dwellings/apartments having and not having access to quiet and urban greenery are analyzed [89–92]. Noneconomic studies on, for example, mental health, focus primarily on restorative aspects [93–95]. Economic valuation studies, like those based on hedonic prices, need to take into account all housing characteristics that are likely to impact the selling price (size, number of bathrooms, etc.), in addition to localization attributes, including availability of green and quiet areas. Urban greenery attributes are, for example, the amount and quality of the trees/greenery on the property [96] or number of nearby trees [97]. Other relevant greenery attributes are distance to parks (quiet areas), size of these areas, suitability, quality, maintenance, litter, and crime [98–103]. Likewise, based on hedonic pricing methodology, statistical techniques could be utilized to extract the relative importance of, for example, acoustical quality and aesthetics for the valuations. However, the value of such regression analyses depends on the availability of suitable indicators of a soundscape's restorative properties, a sufficient number of dwellings (respondents), and sites. To our knowledge, hedonic pricing studies (or other economic valuation studies) comprising more complete soundscape aspects are still lacking.

An alternative economic assessment to hedonic pricing is the stated preference approach, asking people how much they value different aspects of their environment [104] or presenting choice alternatives where environmental quality levels differ between the alternatives [105, 106]. This method has the advantage that it is easier to extract valuations of particular aspects of an environment, such as its perceived restorative properties, for example, by incorporating one or two relevant "willingness to pay" questions in socioacoustic or soundscape research efforts already employing questionnaires.

One possible type of question could elicit the respondents' use of municipal or state funds for increased or decreased availability of restorative areas, changes in how much time is spent, or size of entrance fee deemed acceptable. In most cases, the stated preference methodology is based on extracting individuals' willingness to pay from their own funds for an improvement in some public good quality.

The extracted values are often given as population averages. When applying the values, it may be useful to consider subpopulations and contextual factors. Different groups of users value different qualities of outdoor environments [93, 107, 108]; some soundscapes have cultural, socioeconomic, or personal meaning for their users [109]; and valuations will be coloured by their individual and group experiences. For example, noise-sensitive persons perceive noisy areas to be considerably more annoying than nonsensitive persons [110], and their criteria for what constitutes a restorative area may also be different.

Both hedonic pricing and stated preference approaches focus on the valuation of goods and environmental qualities of the actual users, laypeople, and not on the value assigned by architects, landscape planners, psychologists or health professionals. An adverse soundscape or "noisescape" may also have indirect effects. Noisy areas may restrict social interactions [111] and limit walking and cycling, with potential adverse health consequences [112]. Green areas and other open areas with high-quality soundscapes promote such healthy activities [113–120].

3.6 CONCLUSION

In this chapter, we have explored the notion that access to high-quality acoustic environments may positively affect well-being, quality of life, and environmental health through some restorative or health- and well-being-promoting mechanism. Two types of restoration were discerned: type 1 restoration refers to a high-quality acoustic environment intrinsically providing restoration by way of an immediate pathway; type 2 restoration refers to the effect of availability (knowledge) of a high (better)-quality acoustic environment to a person who otherwise is subject to adverse effects of noise (at home). The latter includes availability of a quiet side in an otherwise high-noise exposure dwelling, or access to or knowledge about nearby green areas. Both can be regarded as providing restoration by way of mediation.

The conclusion, from the little literature available, is that evidence on the intrinsic positive value of areas with high acoustic quality (green areas, wilderness, water) for restoration, by way of an *immediate* pathway, is limited. For restoration by way of *mediation*, studies on the effect of access to a quiet side in the dwelling have established that this results in less noise annoyance at home and also that it has a beneficial effect on sleep quality and blood pressure. There are indications that temporary respite from environmental noise can mitigate the negative effects on health and well-being, but the long-term health effects of this have been little studied. Most studies were cross-sectional, which makes it hard to draw conclusions about the direction of the association.

Different features of the immediate physical environment play together. It has been shown, for example, that access and quality to green space in the immediate vicinity of dwellings moderate the effect of the availability of a quiet side of the dwelling and annoyance. Vice versa, an adverse neighbourhood soundscape (within 75 m) has been shown to modify residential noise annoyance. This effect could be distinguished conceptually and analytically from a lack of quietness. A benefit of quiet neighbourhood areas was not evident in this study, but this could be due to the design not being appropriate for analyzing restorative qualities of quiet areas. Finally, even a view of green space has been shown to moderate the association between road traffic noise and annoyance. Broader studies have shown that a need for quiet space in the wider area is more strongly felt by people who live under noisy conditions and by people who are noise sensitive. Soundscapes perceived negatively are associated with annoyance from all noise sources, a lack of vitality, psychological distress, the prevalence and incidence of depression and anxiety, a higher need for quiet, fewer visits to quiet, and dissatisfaction with access to quiet.

In order to advance our understanding of the process by which these different mechanisms may operate, a more integrated approach is warranted—a contextual one in which the social and physical aspects are studied relating acoustic quality, via restoration, to health, as was outlined by Lercher [121]. Several relevant conceptual models have been found in the literature, each describing part of the mechanism, but a conclusive model has still to be developed. The starting point must be the interaction between a person and his or her environment [10, 28]. Elements from models pertaining to nature and health, and a need for quiet, could subsequently be integrated with state-of-the-art models regarding stress and restoration. Knowledge from studies into the (features in the) built environments that promote restoration may also advance the field. As yet, prerequisite characteristics of urban environments, in order for them to contribute to restoration after stress, are lacking.

In the absence, at present, of health and well-being evidence regarding the restorative value of high-quality soundscapes, a useful approach is to assess their economic value in a holistic manner. Valuation of the availability of quiet and valuation of the availability of restorative areas can be extracted using techniques of economics, whether by hedonic pricing or stated preference methodology. To make these possible, suitable indicators of an area's restorative properties need to be assessed, along with other relevant aspects that enter into the overall assessment.

In Chapter 5, we attempt, based on available data, to derive empirical support for the theoretical approaches described in this chapter.

ACKNOWLEDGEMENTS

In addition to the referenced literature, this chapter is based on conversations and exchanges with many people about the topic. Our special thanks go to Elise van Kempen, Hanneke Kruize, and Mari Murel.

REFERENCES

1. Health Council of the Netherlands (HCN). 1999. Public health impact of large airports. HCN Report 1999/14E. The Hague, September 2. http://www.gr.nl/pdf.php?ID = 19&p = 1.
2. van Kamp, I., Babisch, W., and Brown, A. L. 2012. Environmental noise and health. In *The Praeger Handbook of Environmental Health*, ed. R. H. Friis, chap. 4, ABC-CLIO, LLC, Santa Barbara, CA.
3. De Vos, P., and A. Van Beek. 2011. Environmental noise. In *Encyclopedia of Environmental Health*, ed. J. O. Nriagu, 476–488. Burlington, MA: Elsevier.
4. Payne, S. R., and W. J. Davies. 2009. *Research into the Practical and Policy Applications of Soundscape Concepts and Techniques in Urban Areas.* London: Welsh Assembly Government, Department of the Environment/ Scottish Government, Department for Environment, Food and Rural Affairs.
5. Brown, A. L. 2010. Soundscapes and environmental noise management. *Noise Control Eng. J.*, 58(5): 493–500.
6. Kaplan, R. 1984. Wilderness perception and psychological benefits: An analysis of a continuing program. *Leisure Sci.*, 6: 271–290.
7. Pilcher, E. J., Newman, P., and Manning, R. E. 2008. Understanding and managing experiential aspects of soundscapes at Muir Woods National Monument. *Environ. Manage.*, 43(3): 425–435.
8. Devilee, J., Maris, E., and van Kamp, I. 2010. The societal meaning of sound and noise: A new perspective. Presented at Internoise 2010, Lisbon.
9. Bijsterveld, K. T. 2008. *Mechanical Sound: Technology, Culture, and Public Problems of Noise in the Twentieth Century.* Cambridge, MA: MIT Press.
10. Herranz-Pascual, K., Aspuru, I., and García, I. 2010. Proposed conceptual model of environmental experience as framework to study the soundscape. Presented at Inter Noise 2010: Noise and Sustainability, Lisbon, Paper IN10_445. (Available on CD.)
11. Truax, B., and Barrett, G. W. 2011. Soundscape in a context of acoustic and landscape ecology. *Landsc. Ecol.*, 26(9): 1201–1207.
12. Booi, H., and van den Berg, F. 2012. Quiet areas and the need for quietness in Amsterdam. *Int. J. Environ. Res. Public Health*, 9: 1030–1050.
13. Kang, J. 2011. Noise management: Soundscape approach. In *Encyclopedia of Environmental Health*, ed. J. O. Nriagu, 174–184. Burlington, MA: Elsevier.
14. Brown, A. L., and Muhar, A. 2004. An approach to the acoustic design of outdoor space. *J. Environ. Plan. Manage.*, 47(6): 827–842.
15. Jennings, P., and Cain, R. 2013. A framework for improving urban soundscapes. *Appl. Acoust.*, 74(2): 293–299.
16. Genuit, K., and Fiebig, A. 2006. Psychoacoustics and its benefit for the soundscape approach. *Acta Acust. united Acust.*, 92(6): 952–958.
17. Nilsson, M. E., and Berglund, B. 2006. Soundscape quality in suburban green areas and city parks. *Acta Acust. united Acust.*, 92: 903–911.
18. Lercher, P., and Schulte-Fortkamp, B. 2013. Soundscape of European cities and landscapes: Harmonising. In *Soundscape of European Cities and Landscapes*, ed. J. Kang, K. Chourmouziadou, K. Sakantamis, B. Wang, and Y. Hao. COST. http://noiseabatementsociety.com/wp-content/uploads/2014/01/COST%20 eBOOK%20DOCUMENT%20tryout%2012_5_13%20Final-REVISE19.pdf.

19. Lercher, P., Evans, G. W., and Widmann, U. 2013. The ecological context of soundscapes for children's blood pressure. *J. Acoust. Soc. Am.*, 134(1): 773.

20. Klæboe, R., Kolbenstvedt, M., Fyhri, A., and Solberg, S. 2005. The impact of an adverse neighbourhood soundscape on road traffic noise annoyance. *Acta Acust. united Acust.*, 91(6): 1039–1050.

21. Brown, A. L., Kang J., and Gjestland, T. 2011. Towards standardization in soundscape preference assessment. *Appl. Acoust.*, 72(6): 387–392.

22. Maas, J., Verheij, R. A., Groenewegen, P. P., de Vries, S., and Spreeuwenburg, P. 2006. Green space, urbanity and health: How strong is the relation? *J. Epidemiol. Community Health*, 60: 587–592.

23. Hartig, T., Mang, M., and Evans, G. W. 1991. Restorative effects of natural environmental experiences. *Environ. Behav.*, 23: 3–26.

24. Ottosson, J., and Grahn, P. 2005. A comparison of leisure time spent in a garden with leisure time spent indoors: On measures of restoration in residents in geriatric care. *Landsc. Res.*, 30: 23–55.

25. Rodiek, S. 2002. Influence of an outdoor garden on mood and stress in older persons. *J. Therap. Horticult.*, XIII: 13–21.

26. Kaplan, G. A. 2004. What's wrong with social epidemiology, and how can we make it better? *Epidemiol. Rev.*, 26(1): 124–135. doi: 10.1093/epirev/mxh010.

27. van Kamp, I. 2012. Social aspects of the living environment in relation to environmental health background study. HCN Report 2012/10E. The Hague, July 11.

28. van Kamp, I., Leidelmeijer, Marsman, G., and de Hollander, A. E. M. 2003. Urban environmental quality and human well-being. *Landsc. Urban Plan.*, 65(1–2): 5–8.

29. Murel, M., Kruize, H., van Kamp, I., and van der Sluijs, J. P. 2014. Salutogenic benefits of urban green spaces: A literature review. *Health Place*, under review.

30. Lachowycz, K., and Jones, A. P. 2013. Towards a better understanding of the relationship between greenspace and health: Development of a theoretical framework. *Landsc. Urban Plan.*, 118: 62–69.

31. Health Council of the Netherlands and Council for Spatial Environment and Nature Research, Natuur en gezondheid. 2004. Invloed van natuur op sociaal, psychisch en lichamelijk welbevinden. HCN Report 2004/09, A02a. Den Haag: Gezondheidsraad en RMNO. (In Dutch; English summary.)

32. Mace, B. L., Bell, P. A., and Loomis, R. J. 2004. Visibility and natural quiet in national parks and wilderness areas: Psychological considerations. *Environ. Behav.*, 36(1): 5–31.

33. Krog, N. H., Engdahl, B., and Tambs, K. 2010. Effects of changed aircraft noise exposure on experiential qualities of outdoor recreational areas. *Int. J. Environ. Res. Public Health*, 7(10): 3739–3759.

34. Kaplan, R., and Kaplan, S. 1989. *The Experience of Nature.* Cambridge: Cambridge University Press.

35. Van den Berg, A. E., Koole, S. L. and Van der Wulp, N. Y. 2003. Environmental preference and restoration: (How) are they related? *J. Environ. Psychol.*, 23: 135–146.

36. Hartig, T., and Staats, H. 2003. Guest editors' introduction: Restorative environments. *J. Environ. Psychol.*, 23(2): 103–107.

37. Korpela, K., and Hartig, T. 1996. Restorative qualities of favorite places. *J. Environ. Psychol.*, 16(3): 221–233.

38. Korpela, K. M., et al. 2001. Restorative experience and self-regulation in favorite places. *Environ. Behav.*, 33(4): 572–589.
39. Staats, H. 2003. Understanding pro-environmental attitudes and behavior: An analysis and review of research based on the theory of planned behavior. In *Psychological Theories for Environmental Issues*, ed. M. Bonnes, T. Lee, and M. Bonaiuto, 171–201. Aldershot, UK: Ashgate.
40. Wallenius, M. A. 2004. The interaction of noise stress and personal project stress on subjective health. *J. Environ. Psychol.*, 24: 167–177.
41. Gidlöf-Gunnarsson, A., and Öhrström, E. 2010. Attractive "quiet" court-yards: A potential modifier of urban residents' responses to road traffic noise? *Int. J. Environ. Res. Public Health*, 7: 3359–3375.
42. Gidlöf-Gunnarsson, A., and Öhrström, E. 2007. Noise and well-being in urban residential environments: The potential role of perceived availability of nearby green areas. *Landsc. Urban Plan.*, 83: 115–126.
43. Staats, H., Kieviet, A., and Hartig, T. 2003. Where to recover from attentional fatigue: An expectancy-value analysis of environmental preference. *J. Environ. Psychol.*, 23, 147–157.
44. Brosschot, J. F., Godaert, R., Guido, L., Benschop, R. J., Olff, M., Ballieux, R. E., and Jeijnen, C. J. 1998. Experimental stress and immunological reactivity. A closer look at perceived uncontrollability. *Psychomatic Med.*, 60: 359–361.
45. Van den Berg, A., Hartig, T., and Staats, H. 2007. Preference for nature in urbanized societies: Stress, restoration, and the pursuit of sustainability. *J. Social Issues*, 63, 79–96.
46. Health Council of the Netherlands. 2006. Quiet areas and health. HCN Report 2006/12. The Hague. (In Dutch; English summary.)
47. Hartig, T., Evans, G. W., Jamner, L. D., Davis, D. S., Gärling, T. 2003. Tracking restoration in natural and urban field settings. *J. Environ. Psychol.*, 23: 109–123.
48. Maas, J., Van den Berg, A., Verheij, R. A., and Groenewegen, P. P. 2010. Green space as a buffer between stressful life events and health. *Soc. Sci. Med.*, 70(8): 1203–1210.
49. Karmanov, D., and Hamel, R. 2008. Assessing the restorative potential of contemporary urban environment(s): Beyond the nature versus urban dichotomy. *Landsc. Urban Plan.*, 86(2): 115–125.
50. Kalevi, M., Korepela, K. M., Ylén, M., Tyrväionen, L., Silvennoinen, H. 2010. Favorite green, waterside and urban environments, restorative experiences and perceived health in Finland. *Health Promot. Int.*, daq007.
51. Kaplan, R. 1984. Impact of urban nature: A theoretical analysis. *Urban Ecol.*, 8(3): 189–197.
52. Ulrich, R. S., et al. 1991. Stress recovery during exposure to natural and urban environments. *J. Environ. Psychol.*, 11(3): 201–230.
53. Wells, N. M., and Evans, G. W. 2003. Nearby nature a buffer of life stress among rural children. *Environ. Behav.*, 35(3): 311–330.
54. Van den Berg, A. E., et al. 2010. Green space as a buffer between stressful life events and health. *Social Sci. Med.*, 70(8): 1203–1210.
55. van Kempen, E. E. M. M., and Van Beek, A. J. 2013. De invloed van een stille zijde bij woningen op gezondheid en welbevinden Literatuur en aanbevelingen voor beleid. RIVM Briefrapport 630650005/2013.

56. Forssen, J. 2009. *Road Traffic Noise Levels at Partille Stom after Gap Filling Building Constructions*. Gothenborg, Sweden: Chalmers, University of Technology.

57. Öhrström, E., Skånberg, A., Svensson, H., and Gidlöf-Gunnarsson, A. 2006. *J. Sound Vibr.*, 295(1): 40–59.

58. Selander, J., Nilsson, M. E., Bluhm, G., Rosenlund, M., Lindqvist, M., Nise, G., et al. 2009. Long-term exposure to road traffic noise and myocardial infarction. *Epidemiology*, 20(2): 272–279.

59. De Kluizenaar, Y., Salomons, E. M., Janssen, S. A., van Lenthe, F. J., Vos, H., Zhou, H., ..., Mackenbach, J. P. 2011. Urban road traffic noise and annoyance: The effect of a quiet facade. *J. Acoust. Soc. Am.*, 130(4): 1936–42. doi:10.1121/1.3621180B.

60. De Kluizenaar, Y., Janssen, S., Vos, H., Salomons, E., Zhou, H., and van den Berg, F. 2013. Road traffic noise and annoyance: A quantification of the effect of quiet side exposure at dwellings. *Int. J. Environ. Res. Public Health*, 10(6): 2258–2270. doi: 10.3390/ijerph10062258.

61. Van Renterghem, T., and Botteldooren, D. 2012. Focused study on the quiet side effect in dwellings highly exposed to road traffic noise. *Int. J. Environ. Res. Public Health*, 9(12): 4292–4310. doi: 10.3390/ijerph9124292B.

62. Berglund, B., Lindvall, Th., and Schwela, D. H., eds. 1999. *Guidelines for Community Noise*. Geneva: World Health Organization.

63. Directive 2002/49/EC of the European Parliament and of the Council of 25 June 2002 relating to the assessment and management of environmental noise—Declaration by the Commission in the Conciliation Committee on the directive relating to the assessment and management of environmental noise.

64. Klæboe, R. 2007. Are adverse impacts of neighbourhood noisy areas the flip side of quiet area benefits? *Appl. Acoust.*, 68(5): 557–575.

65. Klæboe, R., Engelien, E., and Steinnes, M. 2006. Context sensitive noise impact mapping. *Appl. Acoust.*, 67(7): 620–642.

66. Klæboe, R., Kolbenstvedt, M., Fyhri, A., and Solberg, S. 2005. The impact of an adverse neighbourhood soundscape on road traffic noise annoyance. *Acta Acust. united Acust.*, 91(6): 1039–1050.

67. Maffiolo, V., Castellengo, M., and Dubois, D. 1999. Qualitative judgements of urban soundscapes. Presented at Proceedings of Inter-Noise, Fort Lauderdale, FL.

68. Li, H. N., Chau, C. K., Tse, M. S., and Tang, S. K. 2012. On the study of the effects of sea views, greenery views and personal characteristics on noise annoyance perception. *J. Acoust. Soc. Am.*, 131(3): 2131–2140.

69. van Kempen, E., van Kamp, I., and Kruize, H. 2011. The need for and access to quiet areas in relation to annoyance, health and sensitivity. In *International Commission on Biological Effects of Noise Proceedings of the 10th International Congress on Noise as a Public Health Problem (ICBEN 2011)*, London, pp. 440–447.

70. van Kamp, I., et al. 2013. Psychological effects, cognitive effects and mental health: Mental health as context rather than health outcome. Presented at Internoise Proceedings, Paper 784.

71. Axelsson, M., Nilsson, E., and Berglund, B. 2010. A principal components model of soundscape perception. *J. Acoust. Soc. Am.*, 128(5): 2836–2846.

72. Schreckenberg, D., Griefahn, B., and Meis, M. 2010. The associations between noise sensitivity, reported physical and mental health, perceived environmental quality, and noise annoyance. *Noise Health*, 12: 7–16.

73. Shepherd, D., Welch, D., Dirks, K. N., and McBride, D. 2013. Do quiet areas afford greater health-related quality of life than noisy areas? *Int. J. Environ. Res. Public Health*, 10: 1284–1303. doi: 10.3390/ijerph10041284.

74. Lercher, P. 2013. Health related quality of life and environmental QoL in soundscape research and implementation. Presented at COST TUD0804, final meeting, Merano, Italy.

75. van Kempen, E., Devilee, J., Swart, W., and van Kamp, I. 2014. Characterizing urban areas with good sound quality: Development of a research design. *Noise Health*, 2014, 16, 73, 380–387.

76. De Coensel, B., and Botteldooren, D. 2006. The quiet rural soundscape and how to characterize it. *Acta Acust. united Acust.*, 92: 887–897.

77. Pheasant, R. J., Fisher, M. N., Watts, G. R., Whitaker, D. J., and Horoshenkov, K. V. 2010. The importance of auditory-visual interaction in the construction of 'tranquil space'. *J. Environ. Psychol.*, 30: 501–509.

78. Pheasant, R. J., Watts, G. R., and Horoshenkov, K. V. 2009. Validation of a tranquility rating prediction tool. *Acta Acust. united Acust.*, 95: 1024–1031.

79. Lindal, P. J., and Hartig, T. 2013. Architectural variation, building height, and the restorative quality of urban residential streetscapes. *J. Environ. Psychol.*, 33: 26–36.

80. Samuelson, P. A. 1954. The pure theory of public expenditure. *Rev. Econ. Stat.*, 387–389.

81. Choumert, J., and Salanié, J. 2008. Provision of urban green spaces: Some insights from economics. *Landsc. Res.*, 33(3): 331–345. doi: 10.1080/01426390802045996.

82. Bickel, P., Friedrich, R., Burgess, A., Fagiani, P., Hunt, A., de Jong, G., Laird, J., Lieb, C., Lindberg, G., and Mackie, P. 2006. Proposal for harmonised guidelines. Deliverable 5, Developing harmonised European approaches for transport costing and project assessment (HEATCO). Stuttgart: Institut für Energiewirtschaft und Rationelle Energieanwendung (IER).

83. Mishan, E. J. 1988. *Cost-Benefit Analysis: An Informal Introduction*. 4th ed. London: Uniwin Hyman.

84. Veisten, K., Smyrnova, Y., Klæboe, R., Hornikx, M., Mosslemi, M., and Kang, J. 2012. Valuation of green walls and green roofs as soundscape measures: Including monetised amenity values together with noise-attenuation values in a cost-benefit analysis of a green wall affecting courtyards. *Int. J. Environ. Res. Public Health*, 9(11): 3770–3778. doi: 10.3390/ijerph9113770.

85. Coley, M. C. 2005. House and landscape value: An application of hedonic pricing: Technique investigation effects of lawn area on house selling price. Master of science thesis, University of Georgia, Athens.

86. Geoghegan, J. 2002. The value of open spaces in residential land use. *Land Use Policy*, 19(1): 91–98.

87. Groff, E., and McCord, E. S. 2011. The role of neighborhood parks as crime generators. *Security J.*, 25: 1–24. doi: 10.1057/sj.2011.1.

88. Luttik, J. 2000. The value of trees, water and open space as reflected by house prices in the Netherlands. *Landsc. Urban Plan.*, 48(3–4): 161–167.

89. Anderson, L. M., and Cordell, H. K. 1985. Residential property values improved by landscaping with trees. *South. J. Appl. Forest.*, 9: 162–166.

90. Anderson, L. M., and Cordell, H. K. 1988. Influence of trees on residential property values in Athens, Georgia (U.S.A.): A survey based on actual sales prices. *Landsc. Urban Plan.*, 15(1–2): 153–164. doi: http://dx.doi.org/10.1016/0169-2046(88)90023-0.

91. Baranzini, A., and Schaerer, C. 2011. A sight for sore eyes: Assessing the value of view and land use in the housing market. *J. Hous. Econ.*, 20(3): 191–199. doi: http://dx.doi.org/10.1016/j.jhe.2011.06.

92. Tyrväinen, L. 1997. The amenity value of the urban forest: An application of the hedonic pricing method. *Landsc. Urban Plan.*, 37(3–4): 211–222. doi: http://dx.doi.org/10.1016/S0169-2046(97)80005-9.

93. Guite, H. F., Clark, C., and Ackrill, G. 2006. The impact of the physical and urban environment on mental well-being. *Public Health*, 120(12): 1117–1126. doi: 10.1016/j.puhe.2006.10.005.

94. Nutsford, D., Pearson, A. L., and Kingham, S. 2013. An ecological study investigating the association between access to urban green space and mental health. *Public Health*, 127(11): 1005–1011. doi: 10.1016/j.puhe.2013.08.016.

95. Sugiyama, T., Leslie, E., Giles-Corti, B., and Owen, N. 2008. Associations of neighbourhood greenness with physical and mental health: Do walking, social coherence and local social interaction explain the relationships? *J. Epidemiol. Community Health*, 62(5): e9.

96. Stigrall, A., and Elam, E. 2009. Impact of improved landscape quality and tree cover on the price of single-family homes. *J. Environ. Horticult.*, 27(1): 24–23.

97. Morales-Fusco, P., Saurí, S., and Lago, A. 2012. Potential freight distribution improvements using motorways of the sea. *J. Transport Geogr.*, 24(0): 1–11. doi: http://dx.doi.org/10.1016/j.jtrangeo.2012.05.007.

98. Andrews, M., and Gatersleben, B. 2010. Variations in perceptions of danger, fear and preference in a simulated natural environment. *J. Environ. Psychol.*, 30: 473–481. doi: 10.1016/j.jenvp.2010.04.001.

99. James, P., Tzoulas, K., Adams, M. D., et al. 2009. Towards an integrated understanding of green space in the European built environment. *Urban Forest. Urban Green.*, 8(2): 65–75. doi: http://dx.doi.org/10.1016/j.ufug.2009.02.001.

100. Kuo, F. E., and Sullivan, W. C. 2001. Environment and crime in the inner city: Does vegetation reduce crime? *Environ. Behav.*, 33: 343–367. doi: 10.1177/00139160121973025.

101. Molnar, B. E., Gortmaker, S. L., Bull, F. C., and Buka, S. L. 2004. Unsafe to play? Neighborhood disorder and lack of safety predict reduced physical activity among urban children and adolescents. *Am. J. Health Promot.*, 18: 378–386. doi: 10.4278/0890-1171-18.5.378.

102. Tzoulas, K., and James, P. 2010. Peoples' use of, and concerns about, green space networks: A case study of Birchwood, Warrington New Town, UK. *Urban Forest. Urban Green.*, 9: 121–128. doi: 10.1016/j.ufug.2009.12.001.

103. Yucel, G. F. 2008. Determination of relationship among demographic variables and the perceptions of safety of urban park users: A case study in three different parks in Istanbul, Turkey. *J. Yasar Univ.*, 3: 1877–1890.

104. Carson, R. T., Flores, N. E., Martin, K. M., and Wright, J. L. 1996. Contingent valuation and revealed preference methodologies: Comparing. *Land Econ.*, 72(1).

105. Ben-Akiva, M. E., and Lerman, S. R. 1985. *Discrete Choice Analysis: Theory and Application to Travel Demand.* Cambridge, MA: MIT Press.

106. Thurstone, L. L. 1927. A law of comparative judgment. *Psychol. Rev.*, 34: 273–286.

107. Richardson, E. A., and Mitchell, R. 2010. Gender differences in relationships between urban green space and health in the United Kingdom. *Social Sci. Med.*, 71: 568–575.

108. Shan, X. Z. 2014. Socio-demographic variation in motives for visiting urban green spaces in a large Chinese city. *Habitat Int.*, 41(0): 114–120.

109. Brambilla, G., and Maffei, L. 2010. Perspective of the soundscape approach as a tool for urban space design. *Noise Control Eng. J.*, 58(5): 532–539. doi: 10.3397/1.3484180.

110. Klæboe, R. 2011. Noise and health: Annoyance and interference. In *Encyclopedia of Environmental Health*, ed. J. O. Nriagu, 152–163. Burlington, MA: Elsevier.

111. Appleyard, D., and Lintell, M. 1972. The environmental quality of city streets: The residents' viewpoint. *AIP J.*, 84–101.

112. Bodin, M., and Hartig, T. 2003. Does the outdoor environment matter for psychological restoration gained through running? *Psychol. Sport Exerc.*, 4(2): 141–153. doi: 10.1016/s1469-0292(01)00038-3.

113. Hillsdon, M., Panter, J., Foster, C., and Jones, A. 2006. The relationship between access and quality of urban green space with population physical activity. *Public Health*, 120(12): 1127–1132. doi: 10.1016/j.puhe.2006.10.007.

114. Kaczynski, A. T., and Henderson, K. A. 2008. Parks and recreation settings and active living: A review of associations with physical activity function and intensity. *J. Phys. Act. Health*, 5(4): 619–632.

115. Ries, A. V., Voorhees, C. C., Roche, K. M., Gittelsohn, J., Yan, A. F., and Astone, N. M. 2009. A quantitative examination of park characteristics related to park use and physical activity among urban youth. *J. Adolesc. Health*, 45(3 Suppl.): S64–S70. doi: 10.1016/j.jadohealth.2009.04.020.

116. Riva, M., Gauvin, L., and Richard, L. 2007. Use of local area facilities for involvement in physical activity in Canada: Insights for developing environmental and policy interventions. *Health Promot. Int.*, 22(3): 227–235. doi: 10.1093/heapro/dam015.

117. Saulle, R., and La Torre, G. 2012. Good quality and available urban green spaces as good quality, health and wellness for human life. *J. Public Health* (Oxf.), 34(1), 161–162. doi: 10.1093/pubmed/fdr090. http://dx.doi.org/10.1016/j.habitatint.2013.07.012.

118. Sugiyama, T., Leslie, E., Giles-Corti, B., and Owen, N. 2008. Associations of neighbourhood greenness with physical and mental health: Do walking, social coherence and local social interaction explain the relationships? *J. Epidemiol. Community Health*, 62(5): e9.

119. Sugiyama, T., Leslie, E., Giles-Corti, B., and Owen, N. 2009. Physical activity for recreation or exercise on neighbourhood streets: Associations with perceived environmental attributes. *Health Place*, 15(4): 1058–1063. doi: 10.1016/j.healthplace.2009.05.001.

120. Takano, T., Nakamura, K., and Watanabe, M. 2002. Urban residential environments and senior citizens' longevity in megacity areas: The importance of walkable green spaces. *J. Epidemiol. Community Health*, 56(12): 913–918.
121. Lercher, P. 2007. Environmental noise: A contextual public health perspective. In *Noise and Its Effects*, ed. L. Luxon and D. Prasher, 345–377. London: Wiley.

ANNEX I: SEARCH STRATEGY I

1. TITLE: soundscape* OR tranquil OR (tranquil-space*) OR (tranquil-landscape*) OR tranquility OR tranquillity OR quiet OR quieter OR (quiet-site) OR (quiet-area*) OR quietness OR (high-acoustics-quality) OR (high-quality-quiet) OR (acoustic-comfort-evaluation) OR (sound-assessment), 5021
2. TITLE-ABS-KEY: environment* OR (public-space*) OR (public-park*) OR (city-park*) OR (green-area*) OR (green-space*) OR urban OR (residential-area*) OR (residential-environment*) OR neighbourhood* OR neighborhood*, 2,753,339
3. #1 AND #2: 710
4. TITLE-ABS-KEY: restoration OR restorative* OR relaxation OR reflection OR wellbeing OR (well-being) OR appealing OR annoyance OR annoying OR comfort OR pleasant* OR (sound-preferences) OR health* OR psychological OR physiological OR physical OR perception* OR perceptive OR perceive* OR perceptual OR attitude* OR opinion*, 7,397,157
5. #3 AND (#4 OR #5), 446
6. (PUBYEAR AFT 1999) AND LANGUAGE(English OR Dutch OR German), 21,925,365
7. #6 AND #7, 385

(TITLE(soundscape* OR tranquil OR (tranquil-space*) OR (tranquil-landscape*) OR tranquility OR tranquillity OR quiet OR quieter OR (quiet-site) OR (quiet-area*) OR quietness OR (high-acoustics-quality) OR (high-quality-quiet) OR (acoustic-comfort-evaluation) OR (sound-assessment))) AND (TITLE-ABS-KEY(environment* OR (public-space*) OR (public-park*) OR (city-park*) OR (green-area*) OR (green-space*) OR urban OR (residential-area*) OR (residential-environment*) OR neighbourhood* OR neighborhood*)) AND (TITLE-ABS-KEY(restoration OR restorative* OR relaxation OR reflection OR wellbeing OR (well-being) OR appealing OR annoyance OR annoying OR comfort OR pleasant* OR (sound-preferences) OR health* OR psychological OR physiological OR physical OR perception* OR perceptive OR perceive* OR perceptual OR attitude* OR opinion*)) OR (TITLE-ABS-KEY(assessment* OR characterization OR characterise OR characteristics OR interpretation OR indicators OR criteria OR parameters

OR labels OR rating OR (perceptual-auditory-attributes) OR quality)) AND ((PUBYEAR > 1999) AND LANGUAGE(English OR Dutch OR German))

Scopus
EXPORT DATE: 10 April 2013

ANNEX II: SEARCH STRATEGY I

Search criteria:

The initial search strategy aimed at retrieving journal papers featuring noise/acoustics/noise annoyance AND restoration AND economic analyses/ cost–benefit analyses. Databases: Science Direct and JASA.

No papers fulfilled all three sets of criteria.

A wider search strategy was subsequently adopted using only two sets of criteria: restoration AND economic analyses/cost–benefit analyses. Databases: Science Direct and JASA + papers previously retrieved by the author on restoration and soundscape.

From this list, papers deemed good examples of how to approach hedonic pricing/stated preference surveys and how to include restorative aspects relevant for economic analyses were selected.

Some references on the economic approach in general were included.

Time period: Some papers published before 2010 relevant for a particular subtopic were included. However, the focus was on recent papers, 2010–early 2014.

Chapter 4

Impact of Soundscape in Terms of Perception

Brigitte Schulte-Fortkamp[1] and André Fiebig[2]
[1]Technische Universität Berlin, Berlin, Germany
[2]Head Acoustics, Herzogenrath, Germany

CONTENTS

4.1 INTRODUCTION

Soundscape exists through human perception of the acoustic environment; this is the first and most important issue when soundscape is considered. The International Organization for Standardization (ISO) provides with the ISO/FDIS 12913-1 2014 a clear definition to understand this innovation in acoustics. Soundscape is an "acoustic environment as perceived or experienced and/or understood by a person or people, in context." Therefore, soundscape research represents a paradigm shift in the field of sound evaluation. First, always, it improves human perception, and second, it expands on classical physical measurements and makes reference to the use of different investigative measurement methods. This multifaceted approach is basic to improving the validity of the research outcome on any subject or phenomenon, and it avoids systematic errors that can occur when relying on only one approach.

The soundscape approach considers the conditions and purposes of a sound's production and how it is perceived. Consequently, it is necessary to understand that the evaluation of sound is a holistic process. ISO/DIS 12913-1 has as its purpose the enabling of a broad international consensus on the definition of *soundscape* and its respective evaluation. It is more than urgent to understand that there is the need to provide a solid foundation for communication across disciplines and professions, with an interest in achievements of better solutions for the people concerned.

4.2 HUMAN PERCEPTION IS THE GUIDANCE

Soundscape as human perception is influenced by the sociocultural background and the psychological dimension with the acoustic environment in context. As a next step in the process, soundscape refers to safety, sustainability, mobility, and ecology (NWIP ISO/CD 12932-2, 2015).

As for the omnipresence of sound in our perceptual world, the sensory element influences the quality of life. The complex interplay between the diverse dimensions leads to classifications of sounds. As for the physical, psychological, psychoacoustic, and cultural dimensions, the classifications might be loudness, pleasantness, disturbance, and comfort. The perception of sound can only be retraced by a multidimensional approach that covers the different dimensions. A common definition of *perception* published in *Business Dictionary* is "the process by which people translate sensory impressions into a coherent and unified view of the world around them." Though necessarily based on incomplete and unverified (or unreliable) information, perception is equated with reality for most practical purposes and guides human behaviour in general. "An environment consists of a finite number of objects, which can be recognized by human beings object by object or as a complex constellation of objects. Based on a number of sensations, which are put together by meditational processes, humans recognize patterns out of sensations. The recognition of patterns is perception" (Fisher et al., 1984).

Concerning soundscape, Figure 4.1 shows a model of constellations in process that will help you understand the dimension of perception.

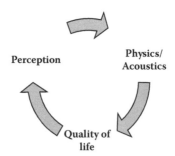

Figure 4.1 Constellations in process. (From Schulte-Fortkamp, B., Soundscape: A matter of human resources, presented at Internoise 2013, Innsbruck, Austria, 2013.)

4.3 SOUNDSCAPE APPROACH

When Schafer in 1977 initiated the soundscape approach, he developed a series of hearing exercises that were aimed at maintaining a high level of sonic awareness. The interaction of people and sound and the way people consciously perceive their environment are therefore central in his approach. His first field studies (World Soundscape Project, founded in the late 1960s) involved level measurements (isobel maps), soundscape recordings, and the description of a wide range of sonic features.

About 40 years later, the soundscape approach is used in different areas and fields. Currently, the so-called soundscape approach and its further development are supported through many research projects. For example, the European COST Action TD 0804 created a network among European soundscaping by integrating soundscape experts not only from Europe, but from all over the world. The ISO/TC 43/SC 1/WG 54 started in 2009 to work on definitions that refer to evaluation procedures (Schulte-Fortkamp and Dubois, 2009a, 2009b, 2009c). Also, soundscape is moving ahead in city planning and supports findings concerning e-mobility and the predicted acceptance by society (Schulte-Fortkamp, 2010a, 2010b). In essence, there is a big change concerning the view of expectation and expertise and its meaning for the development of new products within the innovation of society.

Meanwhile, it is well known that the multidimensional human perception of sound cannot be easily reduced to singular numbers. Moreover, among others, the meaning of sound, the composition of diverse noise sources, the listener's attitude, expectations, and experiences are significant parameters that have to be considered to completely comprehend the different perceptions and evaluations with regard to specific stimuli.

Soundscape suggests exploring noise/sound in its complexity in the respective ambiance. Conditions and purposes of its production, perception, and evaluation have to be considered to understand the evaluation of noise/sound as a holistic approach. Soundscape is understood as an environment of sound with emphasis on the way it is perceived and understood by the individual or by society (see ISO/DIS 12913-1). Everything depends on the relationship between the individual and any such environment. "The soundscape is any acoustic field of study" (Schafer, 1977). Finally accepting the soundscape approach for an evaluation of an acoustic environment will guarantee considering the acoustic environment from a holistic point of view. Therefore, we should talk here about the balance of sounds in respective environments with respect to quality of life. At this point, it becomes obvious why the soundscape approach in his holistic will be the platform to develop methodologies for the evaluation of any acoustic environment. Moreover, it becomes clear why it will guarantee reflection of the acoustical balance in its respective context.

4.4 APPROACHING PEOPLE'S MIND

The soundscape approach requires that perceptual descriptors match physical criteria. Therefore, to fulfil such a requirement, we must correlate language with metrics in engineering. Consequently, it is mandatory to introduce the evaluation methods of psychology and sociology to engineering analysis.

The acoustical properties of sounds and sound quality and their respective assigned meaning and understanding significantly affect the sound evaluation. The context of sound can provoke feelings' sound can be pleasant, familiar, lifestyle representing, helpful for orientation, irritating, and so forth. Respectively, attributes and their meaning have a great impact on the evaluation of the soundscape. The soundscape approach ensures promising access to an acoustic environment through human perception.

Also, improved combined measurement procedures concerning perceptual and physical parameters, including the character of sounds and cross-cultural questionnaires, will be needed. The importance of a survey site selection has to be emphasized. In this case, it comes to product planning: "the product planning tries to take information about those judgments and specify to designers what needs to be accomplished regarding the sound.... A product team therefore has to make the transformation between the judgments of users and the engineering choices for structure, gearing, motors, fans, electrical components, etc., that make up the product. The process needs a mapping between the two" (Lyon, 2000) (Figure 4.2).

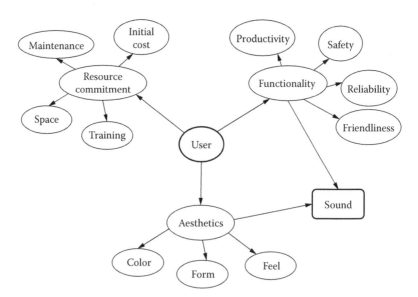

Figure 4.2 Features influencing the judgement of product sound quality. (From Lyon, R. D., *Designing for Product Sound Quality*, New York: Marcel Dekker, 2000.)

In fact, context, cognitive representations, and source identification effect, as well as the sociocultural background of the perception and evaluation of sound have to be considered. Here, further studies in soundscape will help to identify and understand the differences, as well as the similarities, in the perception and evaluation of sound between different areas. An identification of the similarities and differences of sound, dependent on the source, context, activity, and so forth, in situations where, for example, a resident is exposed to environmental noise, will allow a deeper understanding of the evaluation of the acoustic environment.

Soundscape procedures will guarantee the understanding of sounds as meaningful events. Sounds are actually processed and assessed differently by the diversity of cultures and different meanings. Different evaluations can be given to the same acoustical event, depending on living situations.

The challenge for soundscape is to combine human and social sciences with physics within a multidisciplinary investigation process, along with interactive research to improve the quality of life.

4.5 SOUNDSCAPE ANALYSIS

Soundscape analysis will consider and place sound in context linked to activity at realistic study sites. The connection between research and design, for example, for communities, is a creative process. To complete this connection, we need methods to measure and identify (design) values, develop a lexicon of qualities/values for soundscape design, investigate a subject's control/noncontrol over the environment, understand the motivation of people to choose a particular environment, and create soundscape simulations for proposed sites for evaluation by officials.

Soundscape research will provide knowledge that can be applied to create acoustical environments taking into account people's needs. Proposed areas identified for further development have to include economics/noise policy standards, combined effects, common protocols/cross-cultural studies, and education on soundscape, as well as combined measurement procedures concerning perceptual and physical parameters and cross-cultural questionnaires. The importance of a survey site selection has to be emphasized.

The knowledge of assessment, which concerns several noise sources and sensory qualities, must be placed in the context of recent policy developments, which need a firm basis for effective and efficient action. Moreover, multisectorial environmental health impact assessment, the perspective on sustainable development, environmental zoning, citizen involvement, preservation of quiet areas, consideration of "sensitive areas," and the design of "supportive environments" require new insights into the existing annoyance data and new integrative research strategies.

Interviews as well as physical measurement—meaning collecting data with regard to people's expertise in the respective areas—clearly show the evidence of the so-called new experts: the people concerned. Evidently, acceptance and disaffirmation are based on the respective background and strongly related to quality of life. It is important to evaluate each case with respect to such indicators and possible intervention based on people's confirmation. The triangulation of those data is most important.

4.6 PERCEPTION OF SOUND QUALITY

Articles about pleasant and affective sounds often use the term *sound quality* to distinguish their findings from annoyance studies. But, what does sound quality in fact mean? Unfortunately, the term *sound quality* is frequently applied in disparate areas without a theoretical fundament of its concept being provided (Genuit, 2002). A common agreement might be that the term *sound quality* describes the perception of the *adequacy* (Blauert and Jekosch, 1997), *suitability* (Guski, 1997), or *desirability* (Västfjäll and Kleiner, 2002) of a sound attached to the technical object emitting it. However, missing an acknowledged concept of sound quality impedes the profound exploration of the sound quality phenomenon (Fiebig, 2015).

Evidentially, in contrast to psychophysical experiments, the sound quality concept broadens the scope of studying human perception. It comprises an understanding that modifying factors beyond the acoustical stimulus are incorporated or integrated by the listener into their final sound quality judgement (Västfjäll, 2004). However, the semantic concept that is underlying this term varies considerably from study to study. Frequently, sound quality is applied in the context of industrial consumer products (Nykänen, 2008). But, this descriptor is also applied in other sound contexts, like environments (Nilsson and Berglund, 2006), or speech transmission (Lepage et al., 2012). It is obvious that since the perception of sound quality is context dependent, the quality of an acoustical stimulus referring to sound quality cannot be determined without addressing variables like spatial, time, semantic, or response context (Guski and Blauert, 2009). Even in apparently simple psychoacoustic experiments a considerable variability of psychometric functions across laboratories, procedures, and listeners was observed, leading to the insight to properly deal with the influence of context on human multidimensional judgements (Guski and Blauert, 2009).

Sound quality develops when listeners are exposed to an object and judge it with respect to their desires, expectations, and needs in a specific situational context (Genuit and Schulte-Fortkamp, 2010). It evolves from a process in which recognized features are compared to some kind of reference (Blauert and Jekosch, 2007). For example, a consumer has an impression based on a relative evaluation of the sound compared to the expected sound of the product. This comparison is often not based on technical merit or even reality, but is shaped by experiences, where a known highly valued (or high-quality) product sounds a certain way (Pietila, 2013).

Consequently, sound quality perception is based on recognized features that are perceived as desirable or not, which is strongly determined by modifying factors (Västfjäll, 2004). Besides source-related, person-related, and cognitive aspects, Bodden et al. (2000) proposed a *situation-related aspect* in the context of sound quality investigations.

Such contextual factors are responsible for the *relativity of perception* phenomenon, since objects are perceived in dependence of their respective spatial and temporal background (Elfering, 1997). This opens a discussion about the nature of perception, whether absolute perception is possible or human perception is "essentially adaptive in nature" (Helson, 1967). This applies to sound quality perception in particular, which is strongly relative in nature (Zeitler et al., 2006). An approved sound quality today can be rejected tomorrow due to a changed frame of reference. Perception and its communication change over time.

In order to distinguish between the different concepts underlying sound quality studies, Zeitler (2007) proposed a distinction between the *sound character* and *sound quality*. The sound character is related to the basic attributes of an auditory event without considering context, action, and higher level of cognitive processing. The sensation of sound character is ideally devoid of any contextual conditions. Exclusively sensory properties of acoustic events are considered, which could be reliably assessed *without bias* under laboratory conditions (Zeitler et al., 2006). Thus, the sound character concept seems to be more related to auditory sensations and is open to "a parametric representation of a sound" (Sköld et al., 2005). This notion is the basis of psychoacoustics and its indicators connecting the physical world to the perceptual world set to narrow limits. Psychoacoustics describes specific noise perception mechanisms in terms of psychoacoustic parameters, such as loudness, sharpness, roughness, tonality, and fluctuation strength. At least these parameters grasping specific properties of noise can explain more variance in annoyance data than simple sound-pressure-level indicators are able to explain, but since contextual factors also play a role in annoyance, or pleasantness, the variance explanation by means of psychoacoustics still has its limits. Moreover, as pointed out by Marks (1992), even such a simple sensory dimension as loudness cannot be understood as a relatively low-level sensory process, but rather, a process that fundamentally entails the properties of a (higher) cognitive act. Thus, due to complex human cognitive processing of noise stimuli taking into account nonauditory cues as well, in the strict sense, sound character exists, if at all, only in the laboratory. In contrast to it, sound quality describes the perception of sound adequacy affected by diverse factors such as context, cognition, and interaction (Fiebig, 2015). Thus, sound quality cannot be judged without a reference to a concept of expected and desired features of the respective product. Nykänen described this distinction as a suggestion to isolate properties of sound from reactions to sound: a correctly defined sound character should not be dependent on subject or context, whereas sound quality has a higher level of abstraction (Nykänen, 2008). For example, it is obvious that the assessment of music cannot be properly described simply by signal properties. Since affective and emotional reactions are fundamental components of human responses to auditory stimuli (Västfjäll and Kleiner, 2002), the reference to affect and emotion in the context of sound perception might be necessary. For example, Axelssön (2011) worked with the term *affective quality*, referring to the stimulus capacity to change emotional responses, which is not limited to sound stimuli. He also used the term *conceptual quality*, which refers to the meaning of the stimulus. Both terms are only properties of the stimulus and not properties of the perceiver. Jekosch (2005) emphasized from a semiotic perspective that items of perception and items of experience are mutually processed, leading to assignment of meaning to objects. Without this process, objects are meaningless. In contrast to the conceptual distinctions between sound character and sound quality, Blauert and Jekosch (2007) distinguished these terms on the basis of different cognitive levels in a multilayer model applying a bottom-up concept. Product sound quality possesses a higher level of abstraction in the perception process, comprising ideas, concepts, and functions. In contrast to it, the sound quality (as such) describes the lowest abstraction level and determines the quality of sound based on psychoacoustic attributes and form. Moreover, Blauert and Jekosch (2007) defined two intermediate levels: the transmission quality and auditory scene quality.

In this context it is assumed that the acceptance of any sound implicates an act of interpretation (Blauert and Jekosch, 2007). Thus, "the behavior of human beings is not guided by the acoustical signals that we provide with them, ... but rather by the 'meaning' which is transferred via these signals" (Blauert, p. 19, 2005). In the context of responses to music it was shown that "mixed feelings appear to be elicited by complex stimuli with multiple but conflicting affective cues" (Hunter et al., 2008, p. 346).

Furthermore, sound quality is often described as acoustic comfort (Genuit et al., 2005). Zhang et al. (1996) pointed out that comfort and discomfort represent two independent criteria; comfort is often associated with aesthetics, whereas discomfort is more thought to be dependent on physiological or biomechanical factors. They postulated that "comfort and discomfort need to be treated as different and complementary entities in ergonomic investigations" (Zhang et al., 1996, p. 377). Frequently, the orthogonal dimensions arousal (activation) and pleasantness are applied to span the semantic space of sound quality. It requires an internal process of reflecting the perception, matching with internalized expectations, conceptualizing the experiences, and integrating the perceptual results (like feelings and emotions) into expressible categories.

The perception of sound quality is characterized by auditory events that are perceived as pleasant, are easy to identify, support well-being, promote certain activities, and are compatible with nonauditory cues of the corresponding object or environment (Fiebig, 2015).

4.7 PERCEPTION OF ACOUSTICAL ENVIRONMENTS AND SOUNDSCAPES

Humans do not experience the environment by means of sensations; sensations are only the fundament for perceiving the environment. Sensations are interpreted into perceptions, which allow for constructing the environment. This means that humans "construct these perceptions from sensations and from long-term memory of past experiences with similar sensations" (Fischer et al., 1984, p. 19). Thus, a sensation of a soundscape does not exist; it is immediately transformed to perception and the human being cannot disentangle sensation and perception anymore (Fiebig, 2015). Perception includes cognitive processes, since human beings rely on memory of stimulation for comparison with newly experienced stimuli (Fisher et al., 1984). In this context, certain objects are perceived, whereas others are not perceived, depending on attention processes and cognitive processing. Since hearing is the "primary early-warning system" (Scharf, 1998) remaining sensitive to new objects, this sense particularly guides attention.

Considering the acoustical aspects of environments, the mentioned objects can be interpreted as sound sources. The recognition of sound sources is strongly influenced by the perception of the environment, which provides the perceptual frame of reference. As Zeitler (2007) stated, the frame of reference determining perception depends on the actual stimulus context. Axelssön (2011) pointed out that the amount of information must be studied. Amount of information refers to perceptual or conceptual complexity of stimuli, or occurrence of unexpected events. But also, the perceiver's individual capacity to process information, to deal with uncertainty, and to comprehend stimuli plays a significant role in aesthetic appreciation (Axelssön, 2011). This conceptual thinking allows for explaining interindividual preferences, since "people with limited capacity to process information prefer a low amount of information whereas people with a high capacity find reward in the challenge and therefore prefer a high amount of information" (Axelssön, 2011, p. 17). Axelssön's consideration results in the model of information load. Västfjäll and Kleiner (2002) pointed out that in the context of environmental sounds, a broad range of appraisal criteria beyond pleasantness exist, which "modulate the overall sound experience." According to Västfjäll and Kleiner (2002), these factors are related to interpretation and meaning processes.

An essential aspect in the context of the perception of acoustic environments is auditory attention. Attention guides, to a certain extent, how humans perceive and evaluate their environments. Any gained information, which due to mechanisms of attention gets access to working memory, is then evaluated in the working memory, where it can be analyzed, decisions about that information can be made, and plans for action can be elaborated (Knudsen, 2007). Any attention toward stimuli leads to a cognitive benefit, whereas at the same time other stimuli lose importance (cognitive deficit) (Knudsen, 2007). Nykänen (2008) concluded that problems arise when sounds from several sources are analyzed due to the human ability to separate sounds from different sources and consciously or unconsciously focus on some of them. Moreover, it was found that in the presence of numerous sound sources, making it difficult to identify single sources, acoustical sceneries are processed as a whole rather than as independent sound events (Guastavino et al., 2005). In general, it is obvious that the human hearing, in contrast to a sound-level meter, does not work like an absolute measuring instrument, and that humans pay attention to and memorize specific noise patterns and noise structures, instead of taking into account averaged intensities of the sound only (Fiebig, 2013). For example, a higher road traffic noise load can be evaluated as less unpleasant than a low road traffic noise load because of the perceivable single pass-by events penetrating the quietness, particularly attracting attention (Notbohm et al., 2002). Auditory attention processes and source focus mechanisms depend strongly on the number of sound sources and the complexity of the sound situation. Attention processes are influenced by source recognition; as soon as a sound source is recognized, attention is "sharpened," separating source from background and influencing basic auditory sensations like loudness perception (Hellbrück et al., 2004). Any perception of multisource scenarios cannot be understood or predicted in terms of acoustical parameters as long as the source mixture and attention processes are not taken into account. Selective auditory attention processes in perceiving complex (acoustic) environments are still a relevant research subject today. Auditory attention allows humans to focus their mental resources on a particular stream of interest while ignoring others (De Coensel and Botteldooren, 2010). As assumed by Steffens (2013), the need for a selective auditory attention ability regarding focusing on a certain sound source and temporarily ignoring other sound sources has its origin in the evolutionary requirement to recognize potential danger immediately. In this context, an understanding of the main cognitive processing stages is inevitable. Top-down processes have to be distinguished from bottom-up mechanisms. According to De Coensel and Botteldooren (2010), most theories on attention rely on the concept that there is an interplay of bottom-up (saliency-based) and top-down (voluntary) mechanisms in a competitive selection process. Accordingly, Schneider and Parker (2010) concluded that it is apparent that top-down factors and a person's cognitive abilities affect how information is gathered and processed. According to Knudsen (2007), four component processes are fundamental to attention: working memory, competitive selection, top-down sensitivity control, and filtering for stimuli that are likely to be behaviourally important. Due to its relative signal strength, information gains access to working memory by a competitive process, including top-down and bottom-up processing. The information with the greatest signal strength enters the working memory and competes with existing information for control of the working memory (Knudsen, 2007). This concept includes that for improving information quality, sensitivity can be modulated by the top-down mechanism, improving the signal-to-noise ratio in all domains of information processing. At the same time, stimulus-driven access to the working memory could occur (bottom-up attention) when salient stimuli occur infrequently in space or time. These mechanisms selectively gate incoming auditory information-enhancing responses to stimuli that are conspicuous (De Coensel and Botteldooren, 2010). This is accomplished by salience filters, which may also select for stimuli of instinctive or learned biological importance (Knudsen, 2007). Consequently, it can be summarized that a close connection between top-down and bottom-up processes exists. Unexpected or highly salient stimuli can trigger top-down modulations of sensitivity and orienting behaviours (Knudsen, 2007).

A model of attention involved in sound (annoyance) perception is proposed by Andringa and Lanser (2011). They introduce a conceptual model comprising four attentional states: sleep, direct perception, directed attention, and directed attention with a strong sensory distractor. The developed model tries to describe the relations between perception, attention, awareness, attentional control, and motivation in the context of sound annoyance. The different stages of attentional states "become progressively more effortful and less restorative" (Andringa and Lanser, 2011, p. 1). The stressors attract attention involuntarily, and the distracted person tries to diminish or end the stressor.

It is obvious that different attention levels have influences on any sound assessment. This is true for multisource scenarios (Fiebig, 2012), where the attention can shift between different sound sources; for multisensual scenarios, where attention can influence cognitive load (Genell and Västfjäll, 2007); and for simple product tests, where attention is intentionally guided to sound or not (Skoda et al., 2013). Consequently, the consideration of attention processes is essential in order to be able to model human responses to noise and sound perception. However, attention processes are complex and difficult to predict reliably. The probability that attention is paid to a stimulus competing with other stimuli depends on its relative signal strengths. The signal strength is influenced by bottom-up salience filters and is modulated top-down by bias signals that are controlled by working memory (Knudsen, 2007).

Apart from acoustical stimuli, other (environmental) aspects have an impact on the perception of acoustical environments. This means that the frame of reference is constructed on the basis of not only the acoustical stimuli, but also nonacoustical elements, like visual, haptic, olfactory, and gustatory stimuli. The specific interplay of different senses appears to be particularly important for sound perception, which seems to be very susceptible to interference by other senses (Genuit and Fiebig, 2010). For example, Abe et al. (2006) underline the influenceability of sound perception by other sensory sensations with the statement "we unconsciously utilize all sensory information to evaluate sounds." p.51.

In order to reflect the interplay of senses in complex acoustical environments, the soundscape approach gains in importance. The concept of soundscape is not limited to outdoor environments, but is also applied to other contexts, like music halls, metro stations, or passenger cabins. According to Schafer (1977), the concept of soundscape concerns any acoustic field of study, ranging from musical compositions and radio programs to acoustic environments. Because recently the field has evolved differently across disciplines, there is a diversity of opinions about its definition and aims, leading to the phenomenon that the use of the term *soundscape* has become idiosyncratic and ambiguous (ISO/DIS 12913-1, 2013). International standardization endeavours brought forward a common language and understanding of soundscape. Researchers have recently worked with the definition that a soundscape is an "acoustic environment as perceived or experienced and/or understood by a person or people, in context" (ISO/FDIS 12913-1, 2014).

Today, the term *soundscape* is usually applied with respect to (indoor and outdoor) acoustic environments, understanding noise rather as a *resource* than *waste* (Schulte-Fortkamp, 2013). A soundscape is a perceptual construct and must be distinguished from the physical phenomenon—the acoustic environment (Brooks et al., 2014). Moreover, the soundscape perception process is defined as a "continuous, time-varying, and both conscious and unconscious evaluation process by which an individual interprets an acoustic environment in context" (ISO/CD 12913-1, 2010), including physiological, psychological, social, and cultural components. In general, human responses to soundscape are a large set of direct and indirect outcomes (ISO/DIS 12913-1, 2013). Thus, the soundscape approach is complementary to the concept of noise annoyance, which considers the direct outcome of sound perception leading to annoyance based on noise exposure only.

In summary, two major differences exist in the soundscape approach compared to traditional environmental noise assessment and community noise concepts: (1) the same sound perceived in different environments and contexts can lead to different perceptions, experiences, and understandings (interpretations), and (2) sound is not perceived on a continuum ranging from "not at all annoyed" to "extremely annoyed," but is understood as a multidimensional process involving different perceptual dimensions (Fiebig, 2015). Accordingly, a good soundscape quality is not simply identical to the absence of annoyance (Nilsson and Berglund, 2006). Sounds are processed on the basis of semantic features (meaning) rather than abstracted perceptual (sensory) properties. Thus, the cognitive representations of soundscapes have to be explored (Guastavino, 2006).

Axelssön et al. (2010) observed two orthogonal components, *pleasantness* and *eventfulness*, which are involved in soundscape perception. These results support a simple model of soundscape perception based on a small number of basic dimensions, which are related to the informational properties of the soundscapes. Those informational properties of soundscapes, for example, categories of sounds, like technological, natural, or human-made, significantly contribute to soundscape perception, which goes beyond any acoustic descriptor (Axelssön et al., 2010). Andringa and van den Bosch (2013) used the dimensions *pleasure* (valence) and *activation* (arousal), referring to the core affect. However, these dimensions, putting emphasis on emotion, are intimately coupled with the soundscape appraisal based on pleasantness and eventfulness.

Raimbault (2006) stated that different cognitive representations of soundscapes occur, which lead to different human responses to soundscape. A holistic hearing refers to a global representation of an urban soundscape (perception of totality of an acoustic environment), whereas a descriptive listening implies that objects (sound sources) are identified and guide the overall appraisal (Raimbault, 2006).

To study soundscape perception and appraisal, respectively, it is commonly agreed that the expertise of people is needed. As a measurement instrument, local experts have proposed soundwalks for exploring urban areas, opening a field of data for triangulation (Schulte-Fortkamp, 2013). Because of the consideration of context, environment, activity, and voice of the user, it is assumed that high ecological validity in soundscape studies is achieved (Schulte-Fortkamp and Genuit, 2011). Consequently, comparable to traditional environmental noise assessment, it is intended to detect physical noise criteria that match perceptive descriptors, whereby the contrasting difference in the soundscape approach is that it first relies upon human perception and then turns to physical measurement (Brooks et al., 2014). Thus, cognitive, affective, interpretive, and valuative processes are inevitably involved in environmental perception (Fisher et al., 1984).

Basically, the soundscape approach seems to be the most promising in the new field of research in sound design, using the experience of people. These procedures and measurements are proved to lead to novel and effective solutions to change given environments with respect to the acceptance of potential users. Soundscape approaches will be the key issue in future sound design research based on the involvement of people who will guide the research paths. The main challenge is to sufficiently organize the interdisciplinary collaboration. Recent developments regarding soundscape research, such as the ISO/TC 43/SC 1/WG 54 and the COST project TD 0804, will support these promising solutions and connect them in a network.

4.8 CONCEPT OF TRIANGULATION

Evaluation of sound always implies evaluation of lifestyles, depending upon acceptance, and is therefore strongly related to the daily routine of any individual. Moreover, it is highly sensitive to experiences and expectations. Since we need a multifaceted process, triangulation is the leading concept within soundscape.

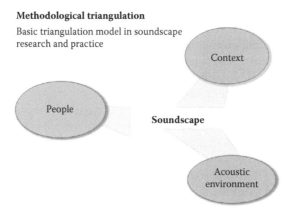

Methodological triangulation

Basic triangulation model in soundscape
research and practice

Figure 4.3 Triangulation. (From De Coensel, B., and Botteldooren, D., A model of saliency-based auditory attention to environmental sound, presented at International Congress on Acoustics 2010, Sydney, Australia, 2010.)

The concept of triangulation is borrowed from navigational and land surveying techniques that determine a single point in space with the convergence of measurements taken from two other distinct points. The idea is that one can be more confident with a result if different methods lead to the same result.

In particular, the application of physical as well as perceptual methods is reasonable. As such, the use of a broad basis of data will provide helpful, accurate results and conclusions regarding the triangulation, which will guarantee a combination of different test results and data layers (Fiebig et al., 2006). Consequently, different approaches should be applied with the intention of achieving a high validity level. With respect to the development of a sound quality metric, subjective ratings, as well as the results of advanced acoustical analyses, are needed. Hence, the application of the latest psychoacoustic scientific findings is required to determine valuable sound quality predictors.

Triangulation for soundscape measurement is a powerful technique that facilitates validation of data through cross-verification of three components: people, context, and acoustic environment. In particular, it refers to the application and combination of several research methodologies in the study of the same phenomenon.

De Coensel and Botteldooren (2010) proposed the model shown in Figure 4.3 with regard to methodological triangulation in soundscapes.

4.9 MODELLING PERCEPTION WITH REGARD TO SOCIETY NEEDS

In summary, it is important to have a consistent framework for defining the context of the soundscape being managed via an interactive approach using the resources through participation and collective management. Relevant stakeholders should be considered, depending on the context. Over the years, strong impact has been given to introduce this concept in science and applications.

While classical noise indicators are known to show strong limitations under certain sound conditions (low-frequency noise, tonal components, multisource environments), it is central to soundscape research and implementation to fit the applied indicators to the perception and appraisal of the concerned people. The fit of indicators depends, however, also on the type of the investigated soundscape. It is extremely important that the fit of indicators reflects the situation and context (personal, social, cultural, land use, economic, geographic) that define the sonic listening space, and also enables tracing dynamic changes, such as time variances of the soundscape, over the day or seasons. In practice, there is still a significant gap between soundscape indicators that are used in some standardized way in measurement by persons and those applied in measurement by instruments. Psychoacoustic, ecological, and landscape acoustics need techniques to be more tightly integrated in such studies to mediate between personal experience and group–area–society requirements and needs.

Only through proper integration of these techniques can the potential of the soundscape approach be implemented in planning and design. The soundscape approach relies, by definition, on this strategy, and in the strict sense it can be said that any study that does not use *triangulation* (people, acoustic environment, context), that is, a combination of several differing investigative methods, cannot be considered a complete soundscape study. So we must look at each soundscape situation from several viewpoints to obtain a more complete picture of the reality. The model on soundscape health relation, including the enviro- and the socio-psycho-physioscape (Lercher and Schulte-Fortkamp, 2013) based on the provided box diagram in ISO/DIS 12913-1 (2013), explains the moderation by context within the soundscape concept with regard to health-related quality of life and adverse health reactions (see Figure 3.2).

4.10 CONCLUSIONS

At present, there is more and more advantage to sufficiently defining the contextual rapport within a given sound evaluation setup and relying on the expertise of the people involved in a soundscape approach: the "new experts" in the field of evaluation. This understanding of evaluation procedures also gives access to new measurement arrangements.

The model by Königstorfer (Figure 4.4) will help you to understand the complexity of decisions with regard to the acceptance of a new approach in established science and applications.

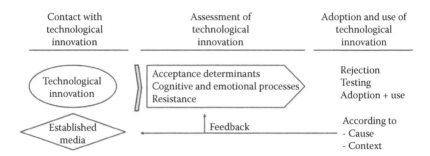

Figure 4.4 Process of acceptance of technological innovation. (Adapted version of a model developed by Königstorfer, T., *Akzeptanz von technologischen Innovationen. Nutzungsentscheidungen von Konsumenten dargestellt am Beispiel von mobilen Internetdiensten*, Wiesbaden: Gabler Verlag, 2008.)

The Environmental Noise Directive 2002/49/EC, related to the assessment and management of environmental noise, demands the preservation of environmental noise quality where it is good. In addition, one major issue of the directive concerns the creation of quiet zones in urban areas. However, it has to be discussed whether the installation and preservation of quiet zones will be the solution with respect to people's concerns in acoustic environments, or whether a general goal should be to discuss acoustically harmonized zones, which will provide a fit to any given area. Creating such areas will be a matter not only for city planners, but also for further stakeholders, such as the industries and people concerned. The ambitious goal of creating acoustic "green" environments cannot be achieved without an interdisciplinary approach, which will be able to holistically reflect the human perception of acoustical sceneries. Sound sources constitute a soundscape, and semantic attributions to sources will become intrinsic properties of the respective acoustic signals.

Psychological and social aspects have a sustainable impact on evaluations and must be considered to create areas that are perceived as harmonic and recreative.

Thus, to understand people's reactions to (environmental) noise, it is imperative to investigate their expectations and experiences besides considering noise metrics and acoustical indicators, which will have an important impact on acceptance of any acoustic environment.

There is a common consent about the necessity of additional parameters besides the A-weighted sound pressure level. Psychoacoustic parameters contribute immensely to measure and assess environmental sound more properly. With the help of psychoacoustic parameters, mainly based on standardized procedures of measurement and analysis, it will be possible to explain some contributors of annoyance caused by environmental noise. But, for the evaluation procedure, contextual and subjective variables have to be integrated. Soundscape is accounting for people's expertise in the respective areas with respect to the acoustical quality of any ambiance (Schulte-Fortkamp and Fiebig, 2006).

"Acoustic design does not, therefore, consist of a set of paradigms. But among others it is an understanding of the balancing mechanism" (Schafer, 1977). The rhythm in soundscape characterizes the time structure of the soundscape with its distinctive noise events embedded in a comprehensive acoustical scenery. "An appreciation of the rhythm is therefore indispensable to the designer who wishes to comprehend how the acoustic environment fits together" (Schafer, 1977). Soundscape can be considered a dynamic system characterized by the time-dependent occurrence of particular sound events embedded in specific environments.

Soundscapes describe ambiances and combine the daily recurrent patterns of multifactorial sound in the process of analysis. An adequate evaluation of environmental noise must reflect the continually varying acoustical scenery and its specific perception in context (Schulte-Fortkamp et al., 2007.) The main message here is that any ambiance, whether it is urban or rural, will be a kind of composition where sounds play an informative role. The information is not based on the loudness level, but on the meaning of the sound.

But, to be successful, the platform for further development, considering resources and ecological input for any given ambiance, has to be defined. Moreover, further emphasis has to be given to sustainable development with regard to the increase of quality of life.

REFERENCES

Abe, K., Ozawa, K., Suzuki, Y., Sone, T. (2006). Comparison of the effects of verbal versus visual information about sound sources on the perception of environmental sounds. *Acta Acust. united Acust.*, 92, 51–60.

Andringa, T.C., Lanser, J.J.L. (2011). Towards causality in sound annoyance. Presented at Internoise 2011, Osaka, Japan.

Andringa, T.C., van den Bosch, K.A. (2013). Core affect and soundscape assessment: Fore- and background soundscape design for quality of life. Presented at Internoise 2013, Innsbruck, Austria.

Axelssön, Ö. (2011). Aesthetic appreciation explicated. Doctoral thesis, Stockholm, Sweden. Stockholm University, Stockholm.

Axelssön, Ö., Nilsson, M.E., Berglund, B. (2010). A principal components model of soundscape perception. *J. Acoust. Soc. Am.*, 128(5), 2836–2846.

Blauert, J. (2005). Analysis and synthesis of auditory scenes. In *Communication Acoustics*, ed. J. Blauert. Berlin: Springer-Verlag, p. 1–20.

Blauert, J., Jekosch, U. (1997). Sound-quality evaluation: A multi-layered problem. *Acta Acust. united Acust.*, 83(5), 747–753.

Blauert, J., Jekosch, U. (2007). Auditory quality of performance spaces for music: The problem of the references. Presented at ICA 2007, Madrid, Spain.

Bodden, M., Heinrichs, R., Blutner, F. (2000). Bewertung von Geräuschqualität in Feld und Labor. Presented at DAGA 2000, Oldenburg, Germany.

Brooks, B., Schulte-Fortkamp, B., Voigt, K.S., Case, A.U. (2014). Exploring our sonic envrionment through soundscape research and theory. *Acoust. Today*, 10(1).

Business Dictionary. http://www.businessdictionary.com/definition/perception.

De Coensel, B., Botteldooren, D. (2010). A model of saliency-based auditory attention to environmental sound. Presented at International Congress on Acoustics 2010, Sydney, Australia.

Elfering, A. (1997). Psychophysikalische Methoden und Ergebnisse in der Bezugs-systemforschung. Die Rolle des Gedächtnisses im Reizgeneralisationsversuch. Doctoral thesis, Frankfurt, Germany.

Fiebig, A. (2012). The link between soundscape perception and attention processes. Presented at Acoustics 2012, Hong Kong, China.

Fiebig, A. (2013). Psychoacoustic evaluation of urban noise. Presented at Internoise 2013, Innsbruck, Austria.

Fiebig, A. (2015). Cognitive stimulus integration in the context of auditory sensations and sound perceptions. Doctoral thesis, Berlin. Dissertation, Berlin, Germany. (Technical University)

Fiebig, A., Schulte-Fortkamp, B., Genuit, K. (2006). New options for the determination of environmental noise quality. Presented at Internoise 2006, Honolulu, HI.

Fisher, J.D., Bell, P.A., Baum, A. (1984). *Environmental Psychology*. New York: CBS College Publishing.

Genell, A., Västfjäll, D. (2007). Vibrations can have both negative and positive effects on the perception of sound. *Int. J. Veh. Noise Vibr.*, 2007, 3(2), 172–184.

Genuit, K. (2002). Sound quality aspects for environmental noise. Presented at Internoise 2002, Dearborn, MI.

Genuit, K., Fiebig, A. (2010). Application of automotive driving simulators for sound and vibration research. *J. Automob. Eng.*, 224(10), 1279–1288.

Genuit, K., Schulte-Fortkamp, B. (2010). Product sound quality and its metrics. Presented at Acoustical Society of Japan Conference, Tokyo, Japan, March 2011.

Genuit, K., Schulte-Fortkamp, B., Fiebig, A. (2005). The acoustical comfort of vehicles: A combination of sound and vibration. Presented at NoiseCon 2005, Minneapolis, MN.

Guastavino, C. (2006). The ideal urban soundscape: Investigating the sound quality of French cities. *Acta Acust. united Acust.*, 92(6), 945–951.

Guastavino, C., Katz, B.F.G., Polack, J.D., Levitin, D.J., Dubois, D. (2005). Ecological validity of soundscape reproduction. *Acta Acust. united Acust.*, 91(2), 333–341.

Guski, R. (1997). Psychological methods for evaluating sound quality and assessing acoustic information. *Acta Acust. united Acust.*, 83(5), 765–774.

Guski, R., Blauert, J. (2009). Psychoacoustics without psychology. Presented at NAG/DAGA 2009, Rotterdam, Netherlands.

Hellbrück, J., Fastl, H., Keller, B. (2004). Does meaning of sound influence loudness judgments? Presented at ICA 2004, Kyoto, Japan.

Helson, H. (1967). Perception. In *Contemporary Approaches to Psychology*, ed. H. Helson and W. Bevan. Princeton, NJ: van Nostrand.

Hunter, P.G., Schellenberg, E.G., Schimmack, U. (2008). Mixed affective responses to music with conflicting cues. *Cognition and Emotion*, 22(29), 327–352.

ISO/DIS 12913-1 (2014). Acoustics. Soundscape Part 1 Definition and conceptual framework. Geneva: International Organization for Standardization. ISO/CD 12932-2 Acoustics—Soundscape—Part 2 Methods and Measurements 2015.

Jekosch, U. (2005). Assigning meaning to sounds: Semiotics in the context of product-sound design. In *Communication Acoustics*, ed. J. Blauert. Berlin: Springer-Verlag.

Knudsen, E.I. (2007). Fundamental components of attention. *Annu. Rev. Neurosci.*, 30, 57–78. doi: 10.1146/annurev.neuro.30.051606.094256.

Königstorfer, T. (2008). *Akzeptanz von technologischen Innovationen. Nutzungsentscheidungen von Konsumenten dargestellt am Beispiel von mobilen Internetdiensten.* Wiesbaden: Gabler Verlag.

Lepage, M., Poschen, S., Kettler, F. (2012). New developments improving speech quality in mobile phones. Presented at DAGA 2012, Darmstadt, Germany.

Lercher, P., Schulte-Fortkamp, B. (2013). Soundscape of European cities and landscapes: Harmonising. In *Soundscape of European Cities and Landscapes*, ed. J. Kang, K. Chourmouziadou, K. Sakantamis, B. Wang, and Y. Hao. COST. http://noiseabatementsociety.com/wp-content/uploads/2014/01/COST%20 eBOOK%20DOCUMENT%20tryout%2012_5_13%20Final-REVISE19.pdf.

Lyon, R.D. (2000). *Designing for Product Sound Quality.* New York: Marcel Dekker.

Marks, L.E. (1992). The slippery context effect in psychophysics: Intensive, extensive, and qualitative continua. *Percept. Psychophys.*, 51(2), 187–198.

Nilsson, M.E., Berglund, B. (2006). Soundscape quality in suburban green areas and city parks. *Acta Acust. united Acust.*, 92(6), 903–911.

Notbohm, G., Gärtner, C., Schwarze, S. (2002). Evaluation of sound quality of vehicle pass-by noises by psycho-physiological measures: Comparison of traffic noise in streets with L- and U-shaped building. Presented at Internoise 2002, Dearborn, MI.

Nykänen, A. (2008). Methods for product sound design. Doctoral thesis, Lulea University of Technology, Sweden.

Pietila, G.M. (2013). Intelligent system approaches to product sound quality analysis. Doctoral thesis, MI. Dissertation, Cincinnati, OH (Univeristy of Cincinnati).

Raimbault, M. (2006). Qualitative judgements of urban soundscapes: Questioning, questionnaires and semantic scales. *Acta Acust. united Acust.*, 92(6), 929–937.

Schafer, R.M. (1977). *The Soundscape: Our Sonic Environment and the Tuning of the World.* Rochester, VT: Destiny Books.

Scharf, B. (1998). Auditory attention, in Pashler, H. (Ed.) *Attention.* Psychology Press, Hove.

Schneider, B.A., Parker, S. (2010). The evolution of psychophysics: From sensation to cognition and back. In *Proceedings of Fechner Day 2010*, vol. 26.

Schulte-Fortkamp, B. (2009a). Subjektive Urteile als objektive Indikatoren für die Beschreibung der neuen "Hybriden." In *Subjektive Fahreindrücke sichtbar machen IV*, ed. K. Becker, 1–7. Expert Verlag. Renningen, Germany.

Schulte-Fortkamp, B. (2009b). The meaning of psychoacoustics for "tomorrow's vehicle." Presented at Proceedings of AAC Aachen Acoustic Colloquium. Aachen, Germany.

Schulte-Fortkamp, B. (2009c). Using the soundscape approach to develop a public space in Berlin. *J. Acoust. Soc. Am.*, 125(5).

Schulte-Fortkamp, B. (2010a). Ökologische Validität und subjektive Evaluation von Geräuschen zur Bestimmung der Qualität von Fahrzeuggeräuschen. In *Sound Engineering*, 121–131. Berlin: Springer Verlag.

Schulte-Fortkamp, B. (2010b). The need for a "green" sound of e-cars: The challenge of passung. Presented at Internoise, Lisbon, Portugal.

Schulte-Fortkamp, B. (2013). Soundscape: A matter of human resources. Presented at Internoise 2013, Innsbruck, Austria.

Schulte-Fortkamp, B., Brooks, B., Bray, W. (2007). Soundscape: An approach to rely on people's perception and expertise in the post-modern community noise era. *Acoust. Today*, 3(1), 7–15.

Schulte-Fortkamp, B., Dubois, D. (eds.). (2006). *Acta Acust. united Acust.*, special issue, *Recent Advances in Soundscape Research*, 92(6).

Schulte-Fortkamp, B., Fiebig, A. (2006). The daily rhythm of soundscape. *J. Acoust. Soc. Am.*, 120(5), 3238.

Schulte-Fortkamp, B., Genuit, K. (2011). Soundscape design and its procedure. Presented at ASJ Conference 2011, Tokyo.

Skoda, S., Steffens, J., Becker-Schweitzer, J. (2013). Investigations on subconscious perception of product sounds. Presented at AIA-DAGA 2013, Merano, Italy.

Sköld, A., Västfjäll, D., Kleiner, M. (2005). Perceived sound character and objective properties of powertrain noise in car compartments. *Acta Acust. united Acust.*, 91(2).

Steffens, J. (2013). Wie viel Realität braucht der Mensch? Untersuchungen zum Einfluss der Versuchsumgebung auf die Geräuschbewertung von Haushaltsgeräten. Doctoral thesis, Düsseldorf, Germany: Technical University of Berlin.

Västfjäll, D. (2004). Contextual influences on sound quality evaluation. *Acta Acust. united Acust.*, 90, 1029–1036.

Västfjäll, D., Kleiner, M. (2002). Emotion in product sound design. Presented at Proceedings of Journées Design Sonore, Paris.

Zeitler, A. (2007). Kognitive Faktoren bei der Skalierung von Höreindrücken. Presented at DAGA 2007, Stuttgart, Germany.

Zeitler, A., Fastl, H., Hellbrück, J., Thoma, G., Ellermeier, W., Zeller, P. (2006). Methodological approaches to investigate the effects of meaning, expectations and context in listening experiments. Presented at Internoise 2006, Honolulu, HI.

Zhang, L., Helander, M.G., Drury, C.G. (1996). Identifying factors of comfort and discomfort in sitting. *Hum. Factors*, 38(3).

Chapter 5

Perceived Soundscapes and Health-Related Quality of Life, Context, Restoration, and Personal Characteristics

Case Studies

Peter Lercher,[1] Irene van Kamp,[2] Eike von Lindern,[3] and Dick Botteldooren[4]

[1]Division of Social Medicine, Medical University Innsbruck (MUI), Innsbruck, Austria

[2]National Institute for Public Health and the Environment, Bilthoven, the Netherlands

[3]Institute for Housing and Urban Research, Uppsala, Sweden

[4]Acoustics Research Group, Ghent University, Ghent, Belgium

CONTENTS

5.1   NEED FOR THE INTEGRATION OF DIFFERENT SCALE LEVELS IN SOUNDSCAPE HEALTH AND QUALITY OF LIFE STUDIES

5.1.1   Introduction: Modelling Exposure or Understanding Soundscapes and Their Negative and Positive Effects on Humans and Their Environment

The assessment of effects on quality of life, annoyance, and health of transport noise at the community level is less straightforward than, for example, for the work site or for other, more closed acoustic spaces or products emitting noise; the results at the community level are much more varied. This often causes administrators and policy makers to wrongly conclude that the evidence for the effects of the acoustic environment on humans is weak. Research into these observed variations in effects provides the key to understanding how adverse effects on health could be mitigated or even prevented by considering health-promoting and restorative aspects of the acoustic environment (healthy soundscapes) in environmental planning and land use assignments. We need, however, to admit that the main current approaches addressing the effects of the acoustic environment on health and quality of life have some inherent methodological limits. Only about 10%–20% of the variance in the community annoyance reactions is explained by typical acoustic indicators (Lden and Lnight) used in regulations. For health effects, the variance explained is even much less (below 5%), and we must ask what is lost and why do we lose essential information on the variance *not* explained. It is obvious that any preventive intervention or implementation of measures at the various scales will suffer from such a deficit. These facts underline the importance of developing approaches and analytic tools in research and practice that improve the predictions, particularly at the specific scales of inquiry and intervention.

Epidemiological studies usually are focused or centred on more severe adverse health effects (e.g., coronary heart disease, hypertension, diabetes) and use standard noise exposure indices (see also Chapter 3). Unfortunately, the accompanying context by which the effects are generally moderated or mediated is not the primary focus of these studies. Generating evidence about average health effects of single sound sources in large populations is the main goal.

Social science surveys on the effects of transport-related noise (e.g., used in environmental health impact assessments) are likewise focused on reporting average distinct psychological and social effects, such as satisfaction/dissatisfaction with the neighbourhood, housing, and the environment; environmental worry and environmental inequality; interference with daily activities; annoyance; and sleep disturbance. The gathered information is then mostly related to classical noise indices and rarely to long-term health outcomes, such as cardiovascular effects or diabetes.

The main limitation of most current *soundscape studies* is that they address perceived sound qualities at a (very) low scale level, such as courtyards, gardens, parks, recreational areas, streets and squares, and other small green urban and suburban areas. Studies mostly employ small samples or use convenience sampling. More critical, too often the necessary last step—to relate the collected perceived soundscape appraisal to health-related quality of life or even defined health effects—is left out (see Figure 3.2 and Chapter 4). Although some soundscape studies use psychoacoustic indicators (details in Chapter 6), surprisingly, often studies use only classical sound indicators—where acoustic information from both approaches would be necessary to appropriately match the obtained human perception and appraisal and provide guidance for both preventive and supportive action.

However, recent conceptual developments in soundscape research attempt to go beyond (see Chapter 4).

The multidisciplinary *Soundscape Support to Health research program* (1999–2007) was the first large effort that focused on the integration of the three approaches outlined above. It resulted in several conceptual (Berglund, 2006) and research articles (Berglund and Nilsson, 2006; Gidlöf-Gunnarsson and Öhrström, 2007, 2010; Öhrström et al., 2006; Skånberg and Öhrström, 2002).

Further conceptual and integrated research input came from the *positive soundscape project* (2006–2009), a multidisciplinary project of five UK universities, including a very wide range of disciplines from sound arts to neurophysiology (Davies et al., 2009a, 2009b, 2013; Hume and Ahtamad, 2013; Jennings and Cain, 2013; Payne, 2013). Our scientific network around the European Cooperation in Science and Technology (COST) project has formulated further requirements and needs (Kang et al., 2013). It did not, however, have research money assigned to implement integrated approaches at medium or larger scales of inquiry.

This chapter aims to fill this gap by providing specific case analyses from studies conducted outside the COST Action TD0804.

5.1.2 Scales, Dimensions, and Facets of the Related Context

5.1.2.1 Scales of Investigation

There is overall consensus in the soundscape community that study information from different scale levels should be collected for the full understanding of the contextual contribution of soundscapes in planning and health prevention and promotion. Based on a COST discussion group in Edinburgh, we derived a typology of soundscape approaches (including their actors) to indicate the three main scales of investigation necessary to reach a sufficiently integrated knowledge base (Figure 5.1). A central role is played by the type II or mid-scale-level investigation, which ideally generates group-level information on all dimensions (soundscape, enviroscape, socioscape, psychophysioscape, and health) when integration of all three types of assessments is enabled by funding agencies.

The second best option is to fund and carry out tandem studies that integrate at least two scale levels (types I and II, types II and III, or types I and III).

5.1.2.2 Dimensions and Facets to Be Studied

The dimensions and facets of the necessary context to be considered cover the wide range schematically visualized in Figure 3.2.

Job and Hatfield (2001) proposed to classify contextual factors related to the community reaction to noise similar to the causal concept of the epidemiological triad (agent–environment–host) as soundscape, enviroscape, and psychscape. In this concept, the enviroscape groups all nonnoise

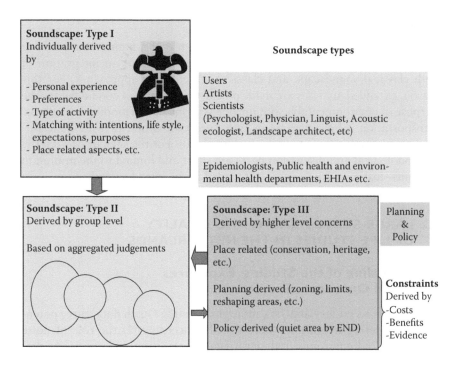

Figure 5.1 Soundscape types of different scales and associated questions, persons, and tasks.

features of the physical environment. In our soundscape block diagram (Figure 3.2), we also include the sociocultural dimension in the enviroscape sphere—like both social geographers and quality of life researchers would do (Marans and Stimson, 2011; Pacione, 2003; Sirgy, 2012).

In Job's classification, host-related factors are subsumed under the term *psychscape*. In our schematic concept (as was described in Figure 3.2), we expand this notion to psycho-physioscape. This term refers to the so-called overall individual adaptive capacity and includes physical abilities and restorative capabilities and links to the World Health Organization (WHO)–European policy framework Health 2020, which aims to establish resilient and restorative environments to support the general adaptive capability and well-being of individual human beings (World Health Organization, 2013).

In addition, theoretical concepts about the potential direct and indirect pathways of action need to be further developed in the field. Authors from related scientific areas (Amérigo and Aragonés, 1997; Astell-Burt et al., 2014; Bonaiuto et al., 1999; Evans and Lepore, 1997; Hartig et al., 2014; Kaplan, 2001; Lachowycz and Jones, 2013; Mitchell et al., 2011; Moser, 2009; Pacione, 2003; Pijanowski et al., 2011; Smith and Pijanowski, 2014; van Kamp et al., 2003; Voigtländer et al., 2014) have provided concepts and ideas to strengthen theory-guided investigations. The application of these concepts will, however, still require adaptation to the aims of the actual study.

The next sections provide insight into the use of a multimethod mix of available analysis options to come closer to the aim of an integration of both classical soundscape and classical noise survey approaches. For this reason, extended analyses are made on existing databases conducted for the purpose of environmental health impact assessment and monitoring of large transportation infrastructures in two European countries with quite different topographic layouts (flat vs. alpine land). For the sake of brevity, the methodological and discussion parts are short and limited to the minimum amount—just to make the points intended in the introductory section.

5.2　LARGE-SCALE HEALTH AND QUALITY OF LIFE STUDIES IN THE NETHERLANDS

5.2.1　Outline of the Studies: Exposures and Outcomes Studied

By means of secondary analysis on several existing Dutch data, three potentially competing hypotheses were tested concerning (1) the role of noise sensitivity and environmental sensitivity, (2) acoustic and environmental quality, and (3) the restorative potential of tranquillity and the need for quiet, especially in people with a mental health problem.

Hypothesis 1: Noise sensitivity hypothesis

    a. Noise sensitivity independently influences annoyance and well-being (additive effect).
    b. Noise sensitivity is a moderator between noise and annoyance (interactive effect).
    c. Noise sensitivity is part of a larger construct of environmental sensitivity.

Hypothesis 2: Environmental sensitivity and sound quality hypothesis
Hypothesis 3: Restoration hypothesis studies 1a and 1b (Schiphol Monitor, 2002, 2005)

5.2.2　Studies 1a and 1b

The two surveys in 2002 and 2005 performed in the framework of the Schiphol Monitor Programme were cross-sectional studies making use of a written questionnaire. The questionnaire informed about health aspects as well of determinants of health, such as age, gender, educational level, and so forth. The surveys can be considered a follow-up of the study performed in 1996 among people living in the vicinity of the airport. The surveys were held 1 year before and 2 years after the opening of a new runway in February 2003 (Houthuijs and van Wiechen, 2006; Breugelmans et al., 2004).

The surveys contained questions about annoyance from a variety of sources, sleep disturbance, whether the participants filed a complaint with the relevant authorities over the past 12 months, disturbance of daily activities, general health, mental health, hypertension, and medication use (sleeping pills, tranquillizers, and antihypertensive medicine). Information on possible confounding factors had already been obtained in the survey of 2002.

Aircraft noise exposure levels were calculated by the National Aerospace Laboratory (NLR), using the Dutch standard calculation model. The NLR delivered annual and biannual exposure levels for aircraft noise (Lden and Lnight) on a 250 × 250 m grid. The x and y coordinates of the residential addresses were linked to the modelled exposure levels using a Geographical Information System (GIS), and the Lden and Lnight values during the last 12 months preceding each questionnaire administration were calculated.

In the questionnaire, single-item 11-point scales (0–10) were used to measure annoyance and sleep disturbance due to aircraft noise (ISO/TS 15666, 2003). Participants scoring 8 or higher were considered to be severely annoyed or sleep disturbed. Results from the first panel round of November 2002 were used to derive dose–response curves for the relationships between noise exposure over the last 12 months and the percentage of severely annoyed, severely sleep disturbed, and those indicating low general health, using the Dutch version of the RAND-36 (van der Zee and Sanderman, 1993).

5.2.3 Studies 2 and 3

Data were derived from a neighbourhood study performed in the Netherlands in 2006 into the association between social, spatial, and physical factors and health/well-being. A *priori*, a selection of 20 neighbourhoods was made with different residential types, ranging from central city, peripheral, and green urban areas to rural, contrasting levels of socioeconomic status (SES) (from highest to lowest status), and the accumulation of environmental exposure noise, air pollution, and access to green space. State-of-the-art cutoff points were used: below and more than 58 dBA accumulated noise), NO_2 concentrations of more than 30 µg/m³, and no green within 500 m from the dwelling. Based on these, every dwelling obtained results in four categories of exceedance (0, 1, 2, 3).

Based on these three variables, a selection was made of three residential types; low versus high SES and low versus high accumulation.

A sample of people of 18 years and older was drawn from the municipal basis registry within the neighbourhood selected. The gross sample was N = 9.502 persons (more than one respondent per address was possible). These people were invited by post to participate, by filling out either a digital or a paper version of the questionnaire. Participants received a voucher worth 5 euros. A reminder was sent twice. The total response rate was 37%, with varying levels of response, with 30% at the lowest end and 51% at the highest end. Per area, 17 people were asked to participate in a nonresponse study that included questions about residential satisfaction, wish to move, gender, age and general health, and level of education.

The questionnaire contained five main clusters: (1) residential situation (dwelling), (2) residential situation (neighbourhood), (3) health and physical activity, (4) well-being and symptoms, and (5) demographics. The mean time for finalization of the questionnaire was 30 minutes.

Questions relevant in the context of this example are the environmental sensitivity scale (Stansfield et al., 1985), quality of life Personal Well-being Index-Adult PWI-A Scale (PWI-A) (Cummins et al., 2010), several subscales of the RAND-36 (van der Zee and Sanderman, 1993), a symptom checklist (Terluin, 1996), and the Utrecht Coping Index (Schreurs et al., 1993).

In a panel study among people living 1 year or less in one of the two selected neighbourhoods, people received a questionnaire in five subsequent years.

5.2.4 Study 4

This study (Baliatsas et al., 2014) combined two data collection methods, a questionnaire survey entitled "Living Environment, Technology and Health" and electronic medical records (EMRs), of adult citizens registered in 21 general practices across the Netherlands. Practices were selected from the primary care database of the Netherlands Institute for Health Services Research (NIVEL). Because the primary focus of the survey was the association between electromagnetic field (EMF) and nonspecific physical symptoms (NSPS), invited participants ($n = 13,007$) were stratified based on preliminary estimates of (low, medium, and high) exposure to mobile phone base stations (Baliatsas et al., 2014). The final number of respondents was $n = 5933$. The privacy regulation of the study was approved by the Dutch Data Protection Authority. More details on the study population and sampling process are described in Kelfkens et al. (2012).

One instrument item on noise sensitivity was used (formatted on a 5-point scale), from a list assessing sensitivity to diverse environmental stressors, adapted from Stansfeld et al. (1985). Participants who reported "strongly agree" on the statement "I am sensitive to noise" formed the highly noise-sensitive (HNS) group. The rest of the sample was considered the low(er)-sensitive (control) (LNS) group.

Similar to the case definition for HNS, respondents who answered "strongly agree" to questions regarding other environmental stressors, such as chemical substances, materials, smells (in general), light, colours, scented detergents, warm or cold environment, temperature changes, and sources of EMFs, were defined as being highly sensitive to these stressors.

To assess NSPS in terms of prevalence, number, and duration, the sum score of 23 items from the Symptoms and Perceptions (SaP) scale (Baliatsas et al., 2014; Yzermans et al., 2015) was used. Higher scores indicate increased symptom number and longer duration. Sleep quality was assessed using a 10-item version of the Groningen Sleep Quality Scale (GSQS) (Meijman et al., 1985; Meijman, 1991); a higher score indicates lower quality of sleep.

Information was obtained on sociodemographic and lifestyle characteristics, such as age, gender, education, ethnic background, home ownership status, degree of urbanization, body mass index (BMI), smoking habits, and alcohol and substance consumption.

Perceived control was assessed based on three items used in Baliatsas et al. (2011): "I am always optimistic about my future," "I hardly ever expect things to go my way," and "If I try, I can influence the quality of my living environment." The score is rated on a 5-point Likert scale, with a higher sum score indicating less perceived control.

Avoidance (coping) behaviour was assessed using the corresponding subscale of the Utrecht Coping List (Schreurs et al., 1993). Items are scored on a 4-point Likert scale; a higher score indicates increased avoidance behaviour, representing the effort to avoid dealing with a stressful situation.

The EMF-related items of the Modern Health Worries (MHW) scale (Petrie et al., 2001; Kaptein et al., 2005) were used to measure participants' levels of concern about potential health effects due to mobile phones, base stations, and high-voltage power lines. Responses were scored on a 5-point Likert scale. A higher sum score indicates higher levels of worry.

5.2.5 Results

5.2.5.1 Noise Sensitivity Hypotheses

What is noise sensitivity? Noise sensitivity (NS) refers to the internal states (physiological, psychological, and attitudinal or related to lifestyle or activities) of individuals that increase their degree of reactivity to noise in general (Job, R. F. S., 1999). Noise sensitivity has a strong genetic component, as was shown by Heinonen-Guzejev et al. (2005). Noise sensitivity can also be caused by physical illness, such as constant migraine headaches, and sudden trauma, such as a head injury. Severe panic disorder may also be accompanied by oversensitive hearing, which in turn facilitates panic attacks. Ear infections, surgery, and the use of some prescribed medications can also lead to this heightened reaction to noise (Heinonen-Guzejev et al., 2005). There is evidence that NS is associated with susceptibility to cardiovascular effects and sleep disturbance (for a review see van Kamp and Davis, 2013). An immediate association with mental health has not been systematically addressed. Exceptions are the studies of Shepherd et al. (2010) and Schreckenberg et al. (2010), who both found NS was associated with health-related quality of life. Noise sensitivity may be a partial indicator of genetically related vulnerability or acquired vulnerability to environmental stressors.

The two first hypotheses have been documented well at an earlier stage; for a review, see van Kamp and Davies (2013). In summary, the findings regarding the role of noise sensitivity, either directly or in interaction with annoyance, are inconclusive. Specifically, the relation with cardiovascular effects needs further attention. An interesting hypothesis has been suggested by Fyhri and Klæboe (2009), who concluded in their paper on noise sensitivity that it is conceivable that individual vulnerability is reflected in both ill health and noise sensitivity.

5.2.5.2 Evidence for Hypothesis Ic

Regarding hypothesis 1c, there is an increasing notion that noise sensitivity is highly correlated with a more general sensitivity for environmental stressors—sometimes referred to as idiopathic environmental intolerance (IEI), but also with a vulnerability to mental health problems. Environmental sensitivity is assumed to have a neurological substrate, being associated with vegetative imbalance as well as mental vulnerability. A study of White et al. (2010) showed that noise sensitivity is significantly associated with environmental sensitivity (0.45) depression (0.32), anger (0.45), fatigue (0.56), tension (0.58), and mental health (0.36), neuroticism (0.52), and inversely with extraversion (−0.36).

Secondary factor analysis on two Dutch data sets revealed a one-component solution in PC analysis of self-reported sensitivity to a range of environmental aspects (Table 5.1). The index used is adapted from Stansfeld et al. (1985).

Table 5.1 PC Analysis on the Environmental Sensitivity Questions in Two Dutch Data Sets (Unpublished)

Component: Environmental Sensitivity	N = 3343 Loading	N = 146 Loading
I immediately notice the colour of things.	0.34	0.40
I am sensitive to smells.	0.54	0.64
Bright light often hurts my eyes.	0.63	0.39
I do not like to touch some materials.	0.63	0.59
I am sensitive to temperature changes.	0.74	0.43
I do not like it when I feel too hot or too cold.	0.68	0.59
I notice immediately when my stomach makes sounds when hungry.	0.49	0.37
I have a low threshold for pain.	0.37	0.31
I am sensitive to noise.	0.63	0.44
I am sensitive to chemical substances.	—	0.73
I am sensitive to radiation.	—	0.55

Internal consistency: sample 1 (0.68) and sample 2 (0.88).

Analyses on several data sets showed that environmental sensitivity was a strong predictor of nonspecific physical symptoms. Multilevel regression analysis on a neighbourhood survey data set in the Netherlands (N = 3343) showed that environmental sensitivity was by far the strongest predictor of nonspecific symptoms (F = 265, p < .0001), over demographic features and SES-related variables, residential satisfaction, and perceived exposure to electromagnetic fields due to base stations (Baliatsas et al., 2011). In a later study (N = 5789) (Baliatsas et al., 2014), the highly environmental sensitive group (N = 514) scored higher on nonspecific physical symptoms in both number and duration than the control group. The most prevalent symptom reported was fatigue. Psychological distress as measured by the 12-item General Health Questionnaire (GHQ-12) and perceived quality of life (Cummins et al., 2010) were strongly associated with the Environmental Sensitivity (ES) Index as well. People scoring high on ES reported higher sick leave (7%) and higher use of painkillers (26%), benzodiazepines (18%), and antidepressants (15%) than the control group. It should be noted that no conclusions can be drawn about causality: it is known that nonspecific physical symptoms are associated with a wide range of somatic and mental illnesses, and it is possible that people with symptoms tend to attribute those to environmental factors.

5.2.5.3 Environmental Sensitivity and Sound Quality Hypothesis (Hypothesis 2)

A few studies position the relationship between noise and mental health in a broader context of environmental quality (Hartig, 2004; Guite et al., 2006). Evidence available is primarily based on the function of green areas and laboratory studies and a few epidemiological studies. These studies address the restorative effects of natural recreational areas outside the urban environment (Rodiek, 2002; Maas et al., 2006; Hartig, 2004; Ottosson, and Grahn, 2005). The role of pleasant sound environments in this process has been understudied. As a consequence, it is not possible to answer the question "What are the prerequisite characteristics for urban environments in order to contribute to restoration after stress?" and likewise, there is little formal policy regulation that addresses these amenities in Europe and elsewhere. Wallenius (2004) have stressed the importance of a focus on daily activities that should provide restoration or demand concentration (e.g., sleeping, relaxing, reading, or studying). Guite et al. (2006) emphasized the importance of social factors in conjunction with physical factors and the need to intervene on both design and social features of residential areas in order to promote mental well-being.

The soundscape approach was originally not oriented toward health and well-being, but toward the meaning of sound in the environment and planning involving the protection and creation of varied soundscapes. However, the concept could potentially link with health through the restorative function of people's experiences of areas of high acoustic quality, and through its emphasis on context and meaning. This approach is still in its infancy, and evidence of the beneficial health effects of areas of high acoustic quality is still lacking, as is the mechanism by which restoration might occur. Moreover, most studies on perceived soundscapes have addressed subjective sound qualities at a (very) low-scale level, such as parks, recreational areas, and squares. Studies on the effects of transport-related noise seldom incorporated perceived soundscapes and are typically focused on negative effects, such as annoyance, sleep disturbance, and environmental worry. It is valuable to know how people describe their sound environment in areas with varying levels of road, air, or rail noise. Available data on perceived soundscapes from the two Schiphol surveys in 2002 ($N = 4255$) and 2005 ($N = 4560$), performed before and after the opening of a new runway, allowed us to perform such analyses. In the context of this chapter, in particular, the interrelations between dimensions of perceived soundscapes and mental health outcomes are relevant. As a start, PC analyses on the perceived soundscape index were performed, revealing two components, which could be interpreted as a negative and a positive dimension (Table 5.2).

The first question addressed was whether noise-level categories for road and air traffic are capable of predicting the scores on the soundscape index. Here too the 2002 Schiphol data were used, the two soundscape scales were dichotomized in a high and a low score (based on the 20% highest vs. 80% lowest), and Lden exposure categories were defined (–52, –57, –62, and –67 for road traffic and –42, –47, –52, –57, –62, and –67 for air traffic). The strength of the association was tested by means of a Wald

Table 5.2 PC Analysis of the Soundscape Questions on the Two Schiphol Data Sets (Unpublished)

Item	2002		2005	
	Loading I	Loading II	Loading I	Loading II
Loud	0.83		0.80	
Aggravating	0.55		0.61	
Sharp	0.73		0.75	
Noisy	0.81		0.79	
Dull	0.58		0.57	
Intruding	0.47		0.80	
Stressful	0.41		0.78	
Disturbing	0.81		0.78	
Light		0.63		0.64
Exciting		0.59		0.54
Calming		0.75		0.78
Pleasant		0.72		0.76

Internal consistency: high for negative scale ($\alpha = 0.88$) and moderate for positive scale ($\alpha = 0.68$).

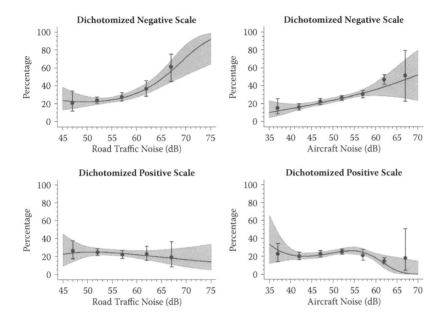

Figure 5.2 Association between categories of road and air traffic levels and the score on the negative and positive soundscape dimension (dichotomized).

test, expressed in the proportion of the variance of negative versus positive soundscape perception explained by noise categories. Figures 5.1 through 5.4 show the direction of the association between noise categories (Lden) and the percentage highly negative/highly positive on the soundscape perception subscales for road and air. Results of the Wald test show a nonlinear significant association between noise categories and the percentage of highly negative perception for road traffic ($F = 6.89$, $p < .000$), and a nonstatistical significant association for the positive dimension. For air traffic noise, the results suggest a decrease in the percentage, with a positive score above 55 dB Lden ($F = 3$, $p < .01$), and a significant linear association between noise categories and the percentage highly negative ($F = 17.26$, $p < .000$).

The high and low groups on the soundscape subscales were subsequently compared on relevant outcomes in the 2002 data only: severe annoyance, vitality, psychological distress, prevalence and incidence of anxiety and depression, the need for quiet (ever and over the past 12 months, and a combination of those), the number of visits to quiet, and dissatisfaction with access to quiet areas. The odds ratios are shown in Table 5.3.

A high score on the negative perceived soundscape scale was associated with severe annoyance from all noise sources, a lack of vitality, psychological distress, the prevalence and incidence of depression and anxiety, a higher need for quiet, less visits to quiet, and dissatisfaction with access to quiet, after adjustment for age, gender, education, ethnicity, and air

Table 5.3 Difference of Means between Low and High Scores
on the Two Perceived Soundscape Scales (Before 2002 Only)

Perception Acoustics Qol	Negative OR (95% CI)	Positive OR (95% CI)
Severe annoyance, all sources	11.0 (6.4–19.1)	0.39 (0.18–0.84)
Vitality (RAND) low	1.6 (1.1–2.3)	0.61 (0.41–0.92)
Psychological distress (GHQ)	2.0 (1.5–2.7)	0.83 (0.61–1.1)
Depression in past 12 months	1.5 (1.1–2.2)	0.76 (0.50–1.2)
Anxiety in past 12 months	1.8 (1.3–2.7)	1.10 (0.70–1.7)
Ever anxious, depressed, worried	1.6 (1.2–2.1)	0.81 (0.61– 1.1)
Anxious, depressed, worried in 12 months	1.7 (1.3–2.4)	0.91 (0.63–1.3)
Need for quiet	2.2 (1.7–3.0)	0.56 (0.42–0.74)
Visit quiet areas	1.7 (1.3–2.3)	0.68 (0.51–0.91)
Dissatisfied access quiet	2.6 (1.9–3.6)	0.45 (0.27–0.76)

Adjusted for age, gender, educational level, ethnicity, exposure categories, and air traffic noise.

traffic noise exposure level. A positive perception was associated with significantly less annoyance from all sources, a higher level of vitality, less need for quiet, and more visits to quiet, and a lower dissatisfaction with access to quiet.

5.2.5.4 *Restoration Hypothesis (Hypothesis 3)*

There is increasing attention for the restorative function of quiet areas where (mental) health effects are concerned (Klæboe, 2005; Gidlöf-Gunnarsson et al., 2010; Gidlöf-Gunnarsson and Öhrström, 2010). There is a clear link with the hypothesis discussed above, but few studies place the relationship between noise and (mental) health explicitly in a broader context of acoustic soundscapes and environmental quality. There is anecdotal evidence that people (especially NS) might profit from a balanced variation of noisy and quiet areas in urban environments (Health Council of the Netherlands, 2006).

Previously reported (Van Kempen et al., 2011) explorative analysis on data from a neighbourhood survey (n = 3364) showed that the score on the mental health scale of the RAND-36, after adjustment for gender, age, Quetelet index (BMI), ethnicity, education, work status, level of urbanization, noise, and noise sensitivity, was associated with a greater need for quiet (χ^2 = 27.5, df = 1, p < .001; OR 0.986, 95% CI 0.980–0.991), more visits to quiet areas (χ^2 = 7.4, df = 1, p = .007; OR 0.993, 95% CI 0.988–0.998), and a lower satisfaction with access to quiet (χ^2 = 11.9, df = 1, p = .001; OR 0.985, 95% CI 0.976–0.993).

5.2.6 Conclusions

We explored a set of potentially competing hypotheses regarding the association between noise and mental health. The associations between environmental sensitivity, perceived soundscapes, and the need for quiet with mental health indicators were respectively analyzed on several different data sets available in the Netherlands. Results showed that environmental sensitivity (IEI), including noise sensitivity, is associated with nonspecific physical symptoms, increased sick leave, and medication use for anxiety and depression, and a lower score on a quality of life index. Secondary analysis showed that perceived soundscapes are also associated with mental health outcomes: a negative perception goes with a higher score on several mental health indicators and higher levels of severe annoyance of all noise sources, while a positive reaction is associated with significantly less noise annoyance and with a higher vitality score on the RAND-36, but not with mental health indicators. Finally, analysis has confirmed that people who perceive their soundscape as more negative have a stronger need for quiet and are less satisfied about the access they have to areas with high acoustic quality. The need for quiet is also confirmed in people who score high on the mental health dimension of the RAND-36. No firm conclusions can be drawn about the direction of these associations. It is well possible that psychological problems (trait or state) make people more susceptible to environmental stimuli, including noise, that they perceive the acoustic quality of their environment as more negative and have a stronger need for places where they can come to rest. In areas where such amenities are not available, it is harder for them to achieve restoration, which in turn might negatively influence their mental health status.

The negative dimension of the soundscape index is highly comparable with annoyance in terms of both association with noise categories and correlational network. The positive dimension is much less associated with noise, or personal characteristics, but possibly more context dependent (amenities, access to green, etc.). A more integral and contextual approach would be needed to unravel the complex and dynamic process of the interaction between people and their environments, in which people with varying susceptibilities are compared. A somewhat similar approach was taken by Crombie et al. (2011) in their analysis on the data from the RANCH study. Results suggest that children with early biological risk—that is, those born prematurely or with a low birth weight—have a greater chance of developing certain mental health outcomes, but are not more vulnerable to the effects of aircraft and road traffic noise at school on mental health. These findings contradict earlier findings (Lercher et al., 2002) of an interaction effect between biological risk and either aircraft or road traffic noise at school on children's mental health. Methodological variations between the two studies might account for these differences.

When studying the mental effects of noise in susceptible groups, two things have to be taken into account: the broader context in terms of available amenities and the differential benefits of these amenities in terms of restoration. This is in line with the conclusions drawn by Schreckenberg et al. (2010), Shepherd et al. (2010) and Lercher (2013).

5.2.7 Discussion of Gains and Limitations

Based on secondary analysis, three main hypotheses were tested regarding the role of personal and contextual variables in health effects of acoustic environments and soundscapes. A previous review showed that studies on the role of NS show inconclusive results. It is well possible that noise sensitivity and ill (mental and physical) health both reflect an underlying vulnerability; there is not necessarily a causal link between noise sensitivity and ill health. Analysis on two data sets showed that self-declared noise sensitivity is moderate to strongly associated with other environmental sensitivities. Environmental sensitivity showed to be a strong predictor of nonspecific symptoms, mental distress, and perceived quality of life. People scoring high on ES reported higher sick leave (7%) and higher use of painkillers (26%), benzodiazepines (18%), and antidepressants (15%). No conclusions can be drawn about causality: it is known that nonspecific physical symptoms are associated with a wide range of somatic and mental illnesses, and it is possible that people with symptoms tend to attribute those to environmental factors.

5.3 MEDIUM- TO SMALLER-SCALE ANALYSES OF SOUND-RELATED HEALTH, QUALITY OF LIFE, AND RESTORATION STUDIES: ENVIRONMENTAL HEALTH IMPACT ASSESSMENT (EHIA) STUDY— BRENNER EISENBAHN GESELLSCHAFT (BEG)

5.3.1 Main Aims and Hypotheses

The studies were conducted within the framework of environmental health impact assessment of a large infrastructure traffic project (rail extensions, rail tunnels). Potential traffic-related adverse impacts on health and quality of life have to be studied in an integrated fashion considering the relevant context and its potential interactions. In environmental health impact assessments, such an approach is required by law and is needed for planning, promoting, and achieving sustainable health-related solutions at the community level.

The conceptual approach underlying these studies extends the classical biomedical and biophysical access by the additional implementation of a theoretical and empirical framework that focuses on human–environment transactions in the tradition of social–ecological environmental stress research (Lercher, 2007) and includes coping with emotional responses to noise (Botteldooren and Lercher, 2004; Lercher, 1996; van Kamp, 1990).

Therefore, we used *broad-based indices* of annoyance, disturbance, and dissatisfaction—including all senses (sound, vibration, air pollution)—and touched also on the emotional affectedness by the overall traffic load to address induced mental distress (Taylor et al., 1997). Job has repeatedly stressed that the commonly used specific noise annoyance questions may not be accurate and powerful enough to detect subtle changes or to elucidate the moderating role of reaction on health outcomes (Job et al., 2001; Job, 1993, 1996). Eventually, the concept addresses health promotional aspects of restoration (Hartig et al., 2003; Kaplan, 2001) and follows a cumulative risk assessment perspective (DeFur et al., 2007; Linder and Sexton, 2011).

5.3.2 Outline of the Studies: Exposures and Outcomes Studied

5.3.2.1 Lower Inn Valley Studies (N = 2004, N = 807, and N = 570)

The study area covered a stretch of about 40 km in the Lower Inn Valley (east of Innsbruck, Austria) and consists of densely populated small towns and villages with a mix of industrial, small business, tourist, and agricultural activities. All studies took place in the same area.

A short summary about study samples is provided in Table 5.4.

The large representative Unterinntal study (UIT-1: N = 2004) was conducted at random by phone (Computer Assisted Telephone Interview [CATI]) in two waves. The first wave was completely at random from the whole valley (including slopes) based on direct (GIS-based) visibility to the traffic lanes. The second wave was directed to sample persons at random near the traffic sources (150 m from highway and railway and 50 m from main roads).

The sampling for the door-to-door interview was based on an a priori GIS stratification of noise exposure (35–44, 45–54, 55–64, >64 *Leq*, dBA) and conducted in a two-step process with replacement. People (aged 18–75 years) were sampled randomly from circular areas around 31 noise measurement sites (radius = 500 m). Persons (807) from 648 households agreed to participate (50.5%) in the survey. The restricted sampling from these circular areas should minimize the known errors of sound propagation predictions for larger distances and increase the validity of the noise assignments. Only persons with a permanent residence of more than 1 year were included in the study. A consecutive survey was conducted to collect more detailed information on the participants' health and the residential environment. Only N = 572 persons agreed to participate in the second wave, which equals a dropout rate of 29.1%. However, no sociodemographic or health-related selection was observed, with the exception of a slightly higher proportion of women compared to the census information. Prior written consent was taken from the participants before the interview, and anthropometric measurements were made.

Table 5.4 Summary Information of the Used Study Data (UIT-I, UIT-2, ALPNAP)

Study, Year, Season	Areas	Traffic Sources	Age Range (years)	N	Methods	Participation	Design/Sampling
EHIA study BEG UIT-I, 1998, summer	32 communities Lower Inn Valley	Motorway, railway, main road	18–75	T1: 1503 T2: 701	Phone: Environment, coping, behaviour, health, medications	81% 83%	Random (based on visibility) Random stratified (based on distance to source)
UIT-II, 1998, fall	Lower Inn Valley	Motorway, railway, main road	18–75	Full: 807 Sub: 572	Interview: Environment Interview: Health, blood pressure, environment, coping behaviour, restoration	51%	Mixed sampling: Stratified by noise from 31 clusters around Measurement points (500 m radius)
ALPNAP study Interreg. IIIB project, 2006, spring	Lower Inn Valley	Motorway, main road, railway	25–74	1653	Phone: Environment, coping, behaviour, health, medications	35% of contacted 45% refused	Two-step stratified random sampling, 4 strata by distance to sources

5.3.2.1.1 Exposure Assessment and Description

The measurement points were selected from two experienced acousticians to cover the variety of topography (valley/slope), settlement structure (housing types, rural/suburban/town), and population density of the area of investigation. The final individual assignment of the source-specific noise exposure (dBA, day and night, *Ldn*) was made after calibration of the modelling results against the measurements from the 31 sites in the centre of the circular areas. All procedures were carried out according to Austrian guidelines (ÖAL No. 28 + 30, ÖNORM S 5011) with a resolution of 25 × 25 m.

In the Alpine Noise and Air Pollution, INTERREG IIIB Alpine Space programme (ALPNAP) study, people were approached by phone (CATI) with a two-step sampling procedure. First, the area address base was stratified by GIS into five sampling groups based on road type and distance. Then persons of the appropriate age range (25–75 years) and gender were randomly sampled with replacement. Overall, 45% did not want to participate, lived for too short of a time in the current home (<1 year), or did not have sufficient hearing or language proficiency. Of those reached on the phone, the individual-level participation was 35%—at the household level it was larger due to women who were more willing to participate (61.5%) than men. There was no other sampling bias with respect to socioeconomic or demographic characteristics. The proportion with good and very good health also did not deviate from the census data—thus, no obvious "healthy attendee" bias (Martikainen et al., 2007; Søgaard et al., 2004) is likely to distort the results.

Two noise exposure calculation methods (MITHRA developed by CSTB in Lyon and an ISO-variant, developed by INTEC in Gent) were used for the individual exposure assignment of the study participants on a 10 × 10 grid basis by GIS. Railway noise emission was extracted from a typical day of noise emission measurements near to the source. An extensive noise-monitoring campaign was conducted to check the validity of these simulations. At 38 locations, so und levels were recorded for more than 1 week during winter (October to January) and during summer (June to August). In addition, the predicted sound pressure levels resulting from PE modelling have been evaluated against these long-term measurements (van Renterghem et al., 2007).

To give insight into the acoustic environment from transportation traffic of this valley, a recording of a typical nightly hour is provided in Figure 5.3. Intermittent rail pass-bys (especially freight trains) strongly dominate during the night. On average, the peak levels of the trains during the night are 13 dBA higher than those from the highway. Also, the mean night level of the rail traffic is nearly 3 dBA higher than during the day. Furthermore, the importance of single pass-bys on the local roads is evident (note the interaction with distance to main roads in the results section).

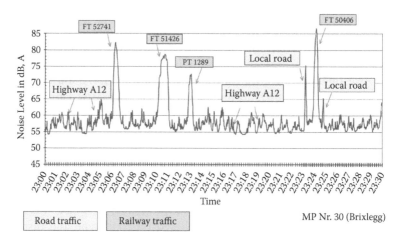

Figure 5.3 Sound levels from transportation during a typical early night hour (23:00–23:30). FT, freight train numbers; PT, personal train numbers.

5.3.3 Results

5.3.3.1 Traffic Noise Exposure, Residential Context, Emotion, and Health: UIT-1 (N = 2004)

In the descriptive analysis of the Eurobarometer questions (43.1 BIS), we observed relevant differences in several dissatisfaction responses between the survey parts T1 and T2 (Table 5.5). Only public transportation was rated more favourable in the T2 sample, where residents on average live closer to the main traffic sources. The largest unfavourable differences in satisfaction were found for area appearance, general living quality, and recreational opportunities. Remarkably, the judgements for general safety and neighbourhood support were also worse near the sources. This illustrates the true width of affectedness (compare with older literature [Appleyard and Lintell, 1972; Jones et al., 1981]) and how deeply the person–environment congruity (Moser, 2009) can actually be disturbed.

As the T2 sample was taken from areas closer to the main sound sources, it is suggestive that differences are due to higher traffic exposures. To explore whether the sound exposure is a significant contributor to the differences, we subjected the five items suggesting a negative effect of traffic to a multivariable regression analysis. For this purpose, we created a five-item scale of sufficient quality (Cronbach's α 0.74). We used the 75th percentile as a cutoff point for a multivariable logistic regression on this dichotomized cumulative area of dissatisfaction outcome.

We found both major sources (highway and railway) significantly contributing to what can be termed *neighbourhood degradation*, showing a nearly linear increase from the lowest to the highest sound levels (Table 5.6). Railway sound levels show a slightly higher effect on high neighbourhood

Table 5.5 Percentage of Responses with High Neighbourhood Dissatisfaction (Eurobarometer): Comparison of Two Areas with Varying Degrees of Traffic Load

"Rather/Very Much Dissatisfied"	Survey	No. of Participants[a]	Percent
Appearance and attractiveness	T1	1414	3.7
	T2	588	13.4
General living quality	T1	1417	3.8
	T2	601	11.1
Recreational opportunities	T1	1397	7.2
	T2	608	18.1
General safety	T1	1404	3.9
	T2	642	6.7
Amount of neighbourhood support	T1	1312	6.2
	T2	617	9.4
Public transportation	T1	1212	27.6
	T2	568	16.0
Personal dissatisfaction with residential area	T1	1376	4.5
	T2	608	9.9

T1 = random sample of the whole area (up to the slopes of the valley), T2 = random sample of persons within 150 m of rail/highway or 50 m of main road. Response in T1 = 81%, response in T2 = 83%.

[a] Missings values vary by type of question.

dissatisfaction. A comparison with the total sound exposure to both sources reveals a slightly higher overall effect for the combined exposure.

The sound level was the second most important contributor. The amount of active coping efforts and the reported emotional response (Figure 5.4a) to traffic were the next strongest predictors, even before self-rated sleep and health status (Figure 5.4b). Scales of environmental vulnerability (sound, air, vibration) and environmental annoyance (sound, air, vibration) remained no longer significant when active coping and emotional response entered the adjusted models. Neither sex nor educational level showed a relevant relationship, while older age was significantly linked with lower neighbourhood dissatisfaction. Apartment housing was significantly related to high dissatisfaction. Home occupation status (as tenant) was associated significantly only in the highway model and showed borderline associations with railway and total sound levels. Overall, the contributors retained the same importance in all sound source models.

5.3.3.2 Traffic Noise Exposure, Disturbance, Coping, Restoration, and Neighbourhood Dissatisfaction: UIT-2 (N = 572)

We followed up our analysis of neighbourhood dissatisfaction with an extended variable set (including two subscales of the General Health Questionnaire [GHQ] and the Perceived Restorativeness Scale [PRS]) from

Table 5.6 Increase in Odds Ratio (95% CI) at Different Sound Levels for Predicted Proportions with High Neighbourhood Dissatisfaction

Sound Source and Neighbourhood Dissatisfaction	Increase in Odds Ratio (95% CI) at Different Sound Levels			Summary Odds Ratio	
Sample UIT-1 (N = 1898)	40–50 Ldn, dBA	50–60 Ldn, dBA	60–70 Ldn, dBA	45–75 Ldn, dBA	Model R²
Total sound level (Ldn, dBA)	1.78 (1.33–2.38)	1.61 (1.41–1.84)	1.50 (1.17–1.91)	3.92 (2.52–6.10)	0.21
Railway sound level (Ldn, dBA)	1.69 (1.30–2.19)	1.48 (1.29–1.69)	1.39 (1.11–1.74)	3.11 (1.99–4.86)	0.20
Highway sound level (Ldn, dBA)	1.45 (1.23–1.71)	1.31 (1.02–1.68)	1.29 (0.95–1.70)	2.26 (1.09–4.68)	0.19

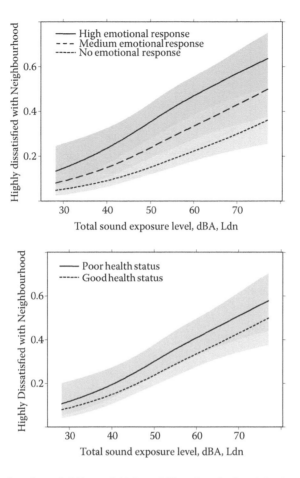

Figure 5.4 Predicted probability of high neighbourhood dissatisfaction with total sound-level exposure by emotional response (a) and by reported health status (b). Models are adjusted for sex, age, education, sleep problems, active coping, home type, and occupation status.

the cluster sample. Note that this cluster sample, collected from the bottom of the valley, experienced *higher exposure levels*, as the clusters are more closely situated to the main traffic sources than the random sample, which also included the slopes at a larger distance from the sources. Therefore, you see higher dissatisfaction at lower sound exposure levels (with moderation by distance to a main road). However, the importance of most variables (age, sex, education, health, home, active coping) remained about the same size. The GHQ score did not contribute more than the simple health status rating and was left off the final model to reduce redundancy. Likewise, the sleep problem scale was not important enough to be kept in the model.

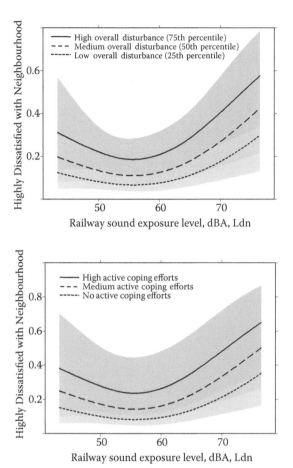

Figure 5.5 Predicted probability of high neighbourhood dissatisfaction with railway sound-level exposure by reported overall disturbance (a) and by amount of active coping efforts (b). Models are adjusted for sex, age, education, home type, health status, fascination (FA) and compatibility (COM) scales of the PRS, distance to the main road, and interaction (IA) distance*railway sound level.

The contribution of the emotional response scale was fully displaced by the dominating contribution of the overall disturbance scale to neighbourhood dissatisfaction in this sample (Figure 5.5a). Therefore, the emotional response variable was excluded in the final model, while active coping kept its significant place (Figure 5.5b). Notably, the second important variable is the compatibility subscale of the PRS, showing an inverse relation with neighbourhood dissatisfaction in spite of substantial sound exposure (Figure 5.6a). In the same direction points the fascination subscale of the PRS. The sound source remains a highly important variable (third place). It shares this importance together with distance to a main road, which

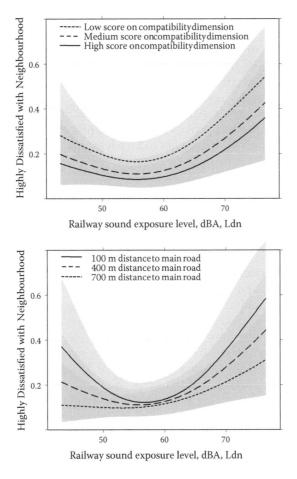

Figure 5.6 Predicted probability of high neighbourhood dissatisfaction with railway sound-level exposure by reported compatibility (a) and observed moderation by distance to a main road (b). Models are adjusted for sex, age, education, home type, health status, overall disturbance, FA and COM scales of the PRS, distance to the main road, and IA distance*railway.

moderates the effect of the larger traffic sources (highway + railway) at lower and higher levels of sound exposure (Figure 5.6b).

5.3.3.3 Integrated Multivariate Model of Traffic Noise Exposure, Residential Satisfaction, and Health: UIT-2 (N = 572)

In this section, one analysis option illustrates how to include the often omitted next or final step to include health as an outcome variable. The health outcome of choice in this analysis is the participant's answer to the simple standard question about the perceived overall actual health status (grades 1–5) during the past 12 months. This measure is a well-known reliable health indicator with a strong potential of predicting further morbidity, health care costs, and even future mortality experience (Benjamins et al., 2004; DeSalvo et al., 2006; Jylhä, 2009; Miilunpalo et al., 1997). To complete the information on the recent health experience (during the past 3 months), the symptom scores of the two subscales (physical and anxiety) were used as additional independent predictors in the full model. Surprisingly, neither sex nor educational level turned out to be relevant factors.

5.3.3.3.1 Results

In the final logistic regression model, both total noise and railway noise exposure are weakly associated with poorer health (grades 3–5) beyond 60 dBA, *Ldn* (Figure 5.7), while highway noise does not show any significant association. However, in the presence of three interactions, the main effect cannot be interpreted correctly. The interactions, including two restoration dimensions (being away and fascination) and emotional affectedness, were significant in the total and railway sound models and showed strong nonlinear components. The highway sound model did only replicate the effect modification through emotional affectedness. Therefore, the true relationship is only represented by the respective sound*interaction curves. The visual result of the interaction concerning emotional affectedness is provided in Figure 5.7b. Nearly identical relationships were observed with railway sound (not shown). Instead, another example (Figure 5.8a) illustrates the shape of the interaction between rail sound and the being away (BA) dimension of the Perceived Restorativeness Scale (PRS) by Hartig.

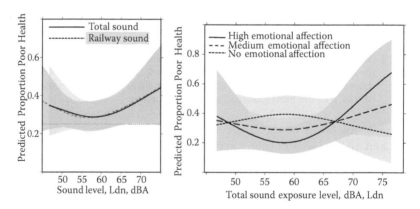

Figure 5.7 Predicted probability of poor health status with total and railway sound-level exposure (a) and observed moderation by emotional affectedness (b). Models are adjusted for age; GHQ score; sensitivity score; neighbourhood satisfaction; FA, BA, and COM scales of the PRS; active and social coping; smoking; sleep score; and IAs sound*FA, sound*BA, and sound*emotion (not (b)).

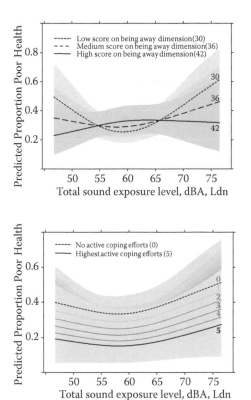

Figure 5.8 Predicted probability of poor health status with railway sound-level exposure by the being away dimension (a) and active coping efforts (b). Models are adjusted for age; GHQ score; sensitivity score; neighbourhood satisfaction; FA, BA (not (a)), and COM scales of the PRS; active (not (b)) and social coping; smoking; sleep score; and IAs sound*FA, sound*BA, and sound*emotion.

A higher score on this restoration dimension is associated with a lower proportion of poor health at all sound levels, while a low score shows an increase at both lower and higher levels of railway sound. Figure 5.8b points to another noteworthy difference in the health model: higher active coping efforts are associated with lower predicted proportions of poor health (= positive effect), while in the neighbourhood dissatisfaction model, an opposite effect has been shown (see Chapter 2 for this issue). A similar positive effect was observed with social coping. Both associations were, however, only significant when tested at the extreme values (low vs. high) and not from low to medium or medium to high scores.

Another notable result is that with average adjustments of model parameters, a relatively high proportion with poorer health is evident over the full range of sound levels. This is mainly due to the fact that age and GHQ symptom score were by far dominant determinants in the model, with

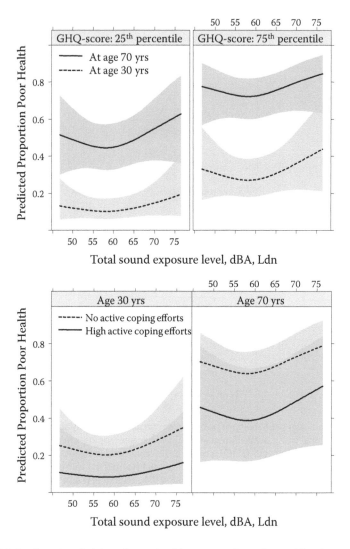

Figure 5.9 Predicted probability of poor health status with total sound-level exposure by the GHQ score and age (a) and total sound level by age and active coping (b). Models are adjusted for GHQ score (not (a)); sensitivity score; neighbour-hood satisfaction; FA, BA, and COM scales of the PRS; active (not (b)) and social coping; smoking; sleep score; and IAs sound*FA, sound*BA, and sound*emotion.

an overall high explanatory power (pseudo R^2 = 0.49). This can easily be observed in the Figure 5.9a and b: at both lower GHQ scores and younger age the proportion with poor health is much lower. We also see that without interactions involved, the weak relationship with the sound level is slightly more pronounced in the healthier groups (low GHQ score/younger age).

The results also make it entirely comprehensible how difficult it is to single out the effects of smaller contextual and psychological determinants in the presence of such dominating contributors (age, GHQ score). Therefore, it is suggested to use further analytic approaches that allow a stronger integration of the contextual variables. The next two examples show such analytic options—one at the aggregate and one at the individual level.

5.3.3.4 Fuzzy Model of Health-Related Quality of Life and Satisfaction: ALPNAP Study (N = 1643)

As preventive interventions to avoid adverse or support positive health need to be implemented at the regional or community level, it is important to know how the actual status of health-related quality of life (HRQoL) varies among communities. In the framework of an EHIA, it is therefore necessary to distinguish between the assessment of the specific satisfaction with the environment, health-related quality of life, subjective well-being (SWB), and general satisfaction with life. Satisfaction with life is considered to be one factor of the wider construct of SWB (Emmons and Diener, 1985), but it does not measure life domains such as health or finances (Diener et al., 1985). Although satisfaction with life is correlated with well-being and health ($r = 0.3$ in this sample), it is more strongly determined by personality, social capital, life circumstances, and economic opportunities (Diener et al., 2013). These factors are not easily amenable to prevention. On the other hand, the health-related quality of life in a certain area is rather determined by the satisfaction with one's physical and mental health and functioning and its relation to the material and psychosocial environment. Thus, more factors from other domains interacting with the immediate environment of the respondent are involved, and a closer examination is needed to distinguish environmental quality of life from personal satisfaction with life.

We applied the Satisfaction with Life Scale (SWLS) and used fuzzy integrals to construct a more complex indicator from 30 basic questions to assess health-related quality of life (HRQoL) at the community level. This fuzzy integration should aid in evaluating the relative usefulness of both indicators to discriminate the HRQoL between the communities. The fuzzy integration technique was earlier applied with annoyance (Botteldooren et al., 2003) and coping (Botteldooren and Lercher, 2004) as criterion variables and later with HRQoL in a different community survey containing the same basic questions (Botteldooren et al., 2006). The responses to each of these questions were aggregated (using Sugeno integral without interaction between factors) to several domains (e.g., stressors, needs, social, work, health), and slightly different weights were applied at each of the aggregation steps for the domains (Table 5.7).

Figure 5.10a and b shows a comparison of the aggregated scores on HRQoL and satisfaction with life by community. While satisfaction with life scores vary only within a range of 0.1, the variation between communities

Table 5.7 Weights Used at Each of the Fuzzy Aggregation Steps for the Health-Related Quality of Life Domains

QoL	Stressors	Needs	Social	Job	Health		
	0.6	0.8	0.8	0.8	—		
Q1 **Stressors**	Air	Noise	Odour	Waste	Landscape	Traffic	Industry
	1	1	—	0.8	0.5	0.8	0.8
Q2 **Needs**	Shopping	Recreation	Schools	Public Transport			
	0.8	0.5	0.5	—			
Q3 **Social**	Appearance	Quality of Living	Security	Neighbourly Support			
	0.2	0.5	0.8	—			
Q28b and Q31 **Stressors at workplace**	Noise	Odour	Heat/Cold	Vibrations	Dirt		
	1	0.5	—	0.8	0.5		
Health	Q32	Q34					
	—	—					

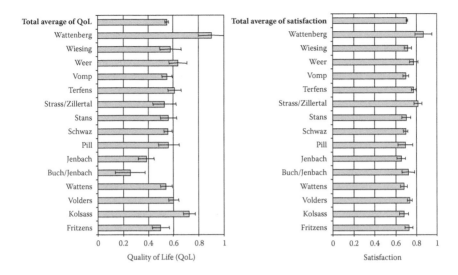

Figure 5.10 Aggregated quality of life (a) and life satisfaction (b) score by community.

shows a much higher range with the results of the fuzzy aggregated health-related quality of life score. This suggests first that quite different determinants influence the two concepts. Second, the quality of life results provide more distinct information about differences at the level, and potential determinants of these community differences can be further investigated to guide interventions at this level. Eventually, the relation of the quality of life scores is stronger, with subjective reports of disturbance by both noise and air pollution, than the measured level of exposure (not shown). This finding further indicates the importance of also measuring other environmental factors both subjectively (e.g., air pollution as perceived odours) and objectively (e.g., air pollution as NO_2 or PM_{10}) and including these as well in soundscape analyses. Notably, measured air pollution was not important in the previous models, although the annual guideline values for NO_2 are exceeded in the area. The result shown in Figure 5.11 suggests that people who have reasons to complain about odourous pollution experience a lower quality of life overall than those complaining about noise. The reasons for this observed difference are not easy to explain and require further analysis.

5.3.3.4.1 Word of Caution and a Summary

We need to remind the reader about the different nature of this aggregate analysis (Greenland, 2001). It is not easy to comprehend why people complaining or not about air pollution generally show lower fuzzy scores on quality of life—as in the previous (regression) analyses on air pollution, we did not observe a relevant effect of air pollution on either neighbourhood

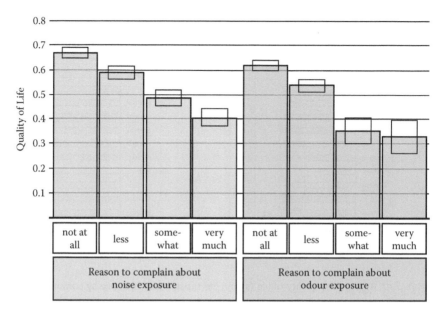

Figure 5.11 Degree of reasons to complain about noise and perceived air pollution (odour).

dissatisfaction or health status. However, with the additional insight from analyses at the aggregate level, the results from descriptive or multiple regression analyses about predictors at the individual level can be counterbalanced (individual- versus aggregate-level bias). Further thoughts are stimulated and suggest hierarchical analyses considering both individual and aggregate data to get more overall confidence for policy recommendations.

5.3.3.5 Integrated SEM of Traffic Noise Exposure, Residential Satisfaction, Restoration, and Health: UIT-2 (N = 572)

Structural equation modelling (SEM) is an alternative integrating statistical technique to test hypotheses using both measured and latent variables and evaluate further the direction and importance of potential relationships with a stronger emphasis on a priori theoretical input (Bollen and Pearl, 2013). SEM has been used in noise, health, and neighbourhood studies for a while (Fyhri and Aasvang, 2010; Hur et al., 2010; Kroesen et al., 2010, 2013; Weden et al., 2008). As we have sufficient *a priori* knowledge, we developed a latent construct including actual exposures (noise, air), susceptibility to traffic (noise, air, vibration), perceived exposures (judgements of exposure severity), and the perceived restorativeness (PRS) of the respondent's home in relation to satisfaction with the living environment (4 items) and health (14 items of GHQ, health status, 5 sleep quality items). The correlations between all latent constructs showed medium to strong associations, and all correlations pointed in the expected direction. We therefore included regression paths between the latent variables in the next step to test our assumptions that health issues, as well as satisfaction with the living environment, are impacted directly and indirectly by the perceived traffic-related exposures and directly by the perceived restorativeness of the living environment (Figure 5.12). We additionally controlled for possible impacts from demographic variables on the latent constructs and regressed each latent construct on the type of housing (single, row, or multiple), gender (male or female), age, education, and density (average people per room). The model fit was acceptably good, although not perfect. The coping construct could not appropriately be accommodated by the model and was therefore excluded. The results, including standardized path coefficients and r^2 values for endogenous latent constructs, are given in Figure 5.12.

5.3.3.5.1 Short Summary Description

Based on the variance explained, we see confirmed a decrease from perceived traffic exposures to residential satisfaction and health, although the final amount of explained variance on health (direct and indirect effects) exceeds the usual size you find with multiple regression analyses. Another important feature is the differential positive effect of the being away dimension of the PRS on the health outcome, while the fascination dimension is (marginally) associated with the residential satisfaction experience. Particularly surprising was the finding that compatibility did not contribute significantly to either health outcomes or residential satisfaction in the SEM. By contrast, compatibility was strongly related to neighbourhood satisfaction in the multiple regression models. Two possible issues may be responsible: (1) the strong correlations found between the PRS dimensions and (2) the different handling and number of variables in the respective models. The relative importance of the restorative function relationships shows the urgent need of the inclusion of a health promotional perspective in both environmental health research and environmental planning considerations. Furthermore, the adjustment variables education level, density, and housing show additional importance not surfacing in the multiple regression models. This analysis again supports the use of alternative statistical methods in terms of a methodological triangulation.

5.3.4 Discussion of Accomplishments and Limitations

5.3.4.1 Accomplishments

- The importance of both the emotional and the perceptional response to the exposures and the significant effects of the restoration dimensions (even at high exposure values).

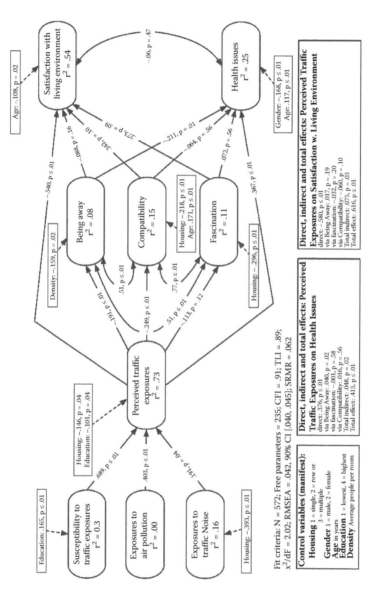

Figure 5.12 Structural equation model with standardized regression path coefficients, explained proportion of variance (r^2), impact of control variables, and calculations for the direct, indirect, and total effect from perceived traffic exposures on both health issues and satisfaction with the living environment. Significant regression paths are bold; only control variables with statistically significant impact are displayed. Manifest indicators for latent constructs and error terms are omitted in the figure due to readability reasons.

- The importance of interactions in the acoustic environment: moderation by sound level and distance to main roads (multisource issue), but also by the emotional response and restoration ability.
- The obvious ambivalent effect of active coping (increasing dissatisfaction, but less adverse impact on health) is another rarely investigated subject.
- The fact that most associations and predictors could be replicated in both the larger and the smaller sample is another feature providing more credibility.
- The large variance explanation of the various models provides further credence into both the analytical results and the potential effects in the case of a planned intervention.
- The analyses implemented at various scale levels can be thought of as a kind of scale triangulation: the applied methodological framework with a span from descriptive, multiple regression, fuzzy aggregation, and structural equation modelling allows a better judgement and evaluation of the obtained results. Eventually, it stimulates further research questions.

5.3.4.2 Limitations

One major limitation of the various case analysis studies is the application of classical sound indicators. Therefore, no comparative assessment of other proposed acoustic and psychoacoustic or multicriteria indicators could be done against both the neighbourhood satisfaction and the health outcome. Another obvious limitation is the cross-sectional design, which prevents a causal interpretation.

In some analyses (UIT-2), the sample size seems at the border of sufficiency (wide confidence limits) to accommodate the relatively large number of adjustment variables and interaction terms. It should, though, be noted that the applied sampling scheme (by prior known noise levels) in the cluster survey resulted in a good distribution of respondents in the middle to higher range of exposure. Conversely, this sampling scheme obviously restricts our statements to middle to higher sound levels. There were too few respondents in the lower sound exposure range. Such information will, however, be needed to make valid statements below 45 dBA, *Ldn*, in order to have a reasonable basis for planning, preserving, and zoning quiet and restorative areas. In particular, the indication of significant moderations by sound events from close-by roads (multisource issue) requires large-scale data below 55 dBA, *Ldn*—what WHO considers reasonable for residential areas. The same applies to the BA restoration dimension, which showed a U-shaped moderated curve—implying that certain groups experience lower restoration associated with poorer health. Nevertheless, the causal path could also be the other way round: persons with poor health have higher expectations (needs) at lower exposure levels and experience less restoration.

5.4 FUTURE PERSPECTIVES

5.4.1 Research Perspective

It is necessary to plan and conduct further soundscape and health research within the larger framework of environmental quality and life quality, restoration, coping, and environmental and social health. The main underlying concepts supporting the understanding of how people respond to distorted soundscapes and what soundscape studies need to address are the stress perspective, the restoration perspective, and the coping perspective.

Future studies need to consider and integrate the research tools, methodologies, and results of neighbouring disciplines dealing with the positive effects of nature and recreation on health (see references and results in Chapter 3).

Therefore, the multiple-stressor concept should be supplemented by its positive counterpart of multiple restoration, where the role of the sonic environment should be tested in the presence of all the other adverse or health-promoting factors.

At the methodological level, properly conducted multilevel analyses can counteract individual-level and aggregate-level bias. Therefore, information should always be gathered from both levels. Tandem study approaches, complementing large representative surveys with smaller-scale intensive surveys or type I studies (Figure 5.1), would accommodate such requirements.

Eventually, the relevant importance of *both* susceptibility and resilience factors (the psychophysioscape) observed in all analyses requires a careful consideration in all future investigations.

5.4.2 Policy Perspective

Multisectoral, integrated environmental health impact assessment, environmental zoning, preservation of quiet areas, sustainable development, consideration of sensitive areas, and the design of supportive and restorative environments are policy developments that require new insights into the interaction between the environment, transportation, and the community. Without this knowledge, the administration will not be able to deal properly with these policy goals at the level of implementation. The current strategy outlined in the Environmental Noise Directive (END) will provide sufficient data to recognize and tackle serious noise problems along major traffic lines and within larger agglomerations. It will fail, however, to respond to noise problems at a medium and a smaller scale, where citizen involvement is an indispensable need and a defensive one-size-fits-all approach does not lead to sustainable solutions (King et al., 2011; Riedel et al., 2013). Note that the soundscape perspective (detailed in Chapters 4 and 6) offers additional planning options to improve the quality of life and also the acoustic quality under unfavourable conditions (see the examples in Chapters 7, 8, and 10 or the EEA soundscape award website of the European Environmental Agency). A successful and sustainable approach requires the integration of the in-depth psychoacoustic and environmental health knowledge by which the results from the perceived soundscape (type I study in Figure 5.1) can be translated into a multisensory (see Chapter 2) policy language (the acoustic soundscape, the visual, the perceptual, and the social environment) that is also effective at the larger scale. Only the combined consideration of this recently accumulated advanced knowledge in planning, architecture, and land use will guarantee the implementation of a preventive and population-promoting perspective rooted in sustainability, restoration, coping abilities, quality of life, and health.

REFERENCES

Amérigo, M., and Aragonés, J. I. (1997). A theoretical and methodological approach to the study of residential satisfaction. *Journal of Environmental Psychology*, 17(1), 47–57. doi: 10.1006/jevp.1996.0038.

Appleyard, D., and Lintell, M. (1972). The environmental quality of city streets: The residents' viewpoint. *Journal of the American Institute of Planners*, 38(2), 84–101. doi: 10.1080/01944367208977410.

Astell-Burt, T., Mitchell, R., and Hartig, T. (2014). The association between green space and mental health varies across the lifecourse. A longitudinal study. *Journal of Epidemiology and Community Health*, jech-2013-203767. doi: 10.1136/jech-2013-203767.

Baliatsas, C., van Kamp, I., Hooiveld, M., Yzermans, J., and Lebret, E. (2014). Comparing non-specific physical symptoms in environmentally sensitive patients: Prevalence, duration, functional status and illness behavior. *Journal of Psychosomatic Research*, 76(5), 405–413.

Baliatsas, C., van Kamp, I., Kelfkens, G., Schipper, M., Bolte, J., Yzermans, J., and Lebret, E. (2011). Non-specific physical symptoms in relation to actual and perceived proximity to mobile phone base stations and powerlines. *BMC Public Health*, 11, 421.

Benjamins, M. R., Hummer, R. A., Eberstein, I. W., and Nam, C. B. (2004). Self-reported health and adult mortality risk: An analysis of cause-specific mortality. *Social Science and Medicine*, 59(6), 1297–306. doi: 10.1016/j.socscimed.2003.01.001.

Berglund, B. (2006). From the WHO guidelines for community noise to healthy soundscapes. In *Proceedings of the Institute of Acoustics 2006*, pp. 1–8.

Berglund, B., and Nilsson, M. E. (2006). On a tool for measuring soundscape quality in urban residential areas. *Acta Acustica united with Acustica*, 92(6), 938–944.

Bollen, K. A., and Pearl, J. (2013). Eight myths about causality and structural equation models. In *Handbook of Causal Analysis for Social Research*, ed. S. L. Morgan, 301–328. Dordrecht: Springer. doi: 10.1007/978-94-007-6094-3.

Bonaiuto, M., Aiello, A., Perugini, M., Bonnes, M., and Ercolani, A. (1999). Multidimensional perception of residential environment quality and neighbourhood attachment in the urban environment. *Journal of Environmental Psychology*, 19(4), 331–352. doi: 10.1006/jevp.1999.0138.

Botteldooren, D., and Lercher, P. (2004). Soft-computing base analyses of the relationship between annoyance and coping with noise and odor. *Journal of the Acoustical Society of America*, 115(6), 2974–2985.

Botteldooren, D., Verkeyn, A., de Baets, B., and Lercher, P. (2006). Fuzzy integrals as a tool for obtaining an indicator for quality of life. In *Proceedings of the 2006 IEEE World Congress on Computational Intelligence*, Vancouver, pp. 5985–5991.

Botteldooren, D., Verkeyn, A., and Lercher, P. (2003). A fuzzy rule based framework for noise annoyance modeling. *Journal of the Acoustical Society of America*, 114(3), 1487–1498.

Breugelmans, O. R. P., Heisterkamp, S. H., and Houthuijs, D. J. M. (2004). Gezondheid en beleving van de omgevingskwaliteit in de regio Schiphol: 2002-Tussenrapportage Monitoring Gezondheidskundige Evaluatie Schiphol. RIVM Rapport 630100001, Netherland.

Crombie, R., Clark, C., and Stansfeld, S. A. (2011). Environmental noise exposure, early biological risk and mental health in nine to ten year old children: a cross-sectional field study. *Environmental Health: A Global Access Science Source*, 10, 39. doi:10.1186/1476–069X–10–39.

Cummins, R. A., Mellor, D., Stokes, M. A., and Lau, A. L. D. (2010). Measures of subjective wellbeing. In *Rehabilitation and Health Assessment: Applying ICF Guidelines*, ed. E. Mpofu and T. Oakland, 409–423. New York: Springer.

Davies, B., Cain, R., Adams, M., Bruce, N., Carlyle, A., Cusack, P., Hume, Ken., Jennings, P. A., and Plack, C. J. (2009a). A positive soundscape evaluation tool. In Proceedings of the 8th European Conference on Noise Control-Abstracts. St. Albans, Institute of Acoustics, UK.

Davies, W. J., Adams, M. D., Bruce, N. S., Cain, R., Carlyle, A., Cusack, P., Hall, D. A., Hume, K. I., Irwin, A., Jennings, P., Marselle, M., Plack, C. J., and Poxon, J. (2013). Perception of soundscapes: An interdisciplinary approach. *Applied Acoustics*, 74(2), 224–231. doi: 10.1016/j.apacoust.2012.05.010.

Davies, W. J., Adams, M. D., Bruce, N. S., Carlyle, A., and Cusack, P. (2009a). A positive soundscape evaluation tool. Changes.

Davies, W. J., Adams, M. D., Bruce, N. S., Marselle, M., Cain, R., Jennings, P., Poxon, J., Carlyle, A., Cusack, P., Hall, D. A., Irwin, A., Hume, K., and Plack, C. J. (2009b). The positive soundscape project: A synthesis of results from many disciplines. Proceedings of Internoise 2009. Ottawa, Institute of Noise Control Engineering (INCE), Canada. Presented at Internoise 2009. Proc. of Internoise 2010, Lissabon, Soc. Portuguesa de Acustica, Portugal.

DeFur, P. L., Evans, G. W., Cohen Hubal, E. A., Kyle, A. D., Morello-Frosch, R. A., and Williams, D. R. (2007). Vulnerability as a function of individual and group resources in cumulative risk assessment. *Environmental Health Perspectives*, 115(5), 817–24. doi: 10.1289/ehp.9332.

DeSalvo, K. B., Bloser, N., Reynolds, K., He, J., and Muntner, P. (2006). Mortality prediction with a single general self-rated health question: A meta-analysis. *Journal of General Internal Medicine*, 21(3), 267–75. doi: 10.1111/j.1525-1497.2005.00291.x.

Diener, E., Emmons, R. A., Larsen, R. J., and Griffin, S. (1985). The Satisfaction with Life Scale. *Journal of Personality Assessment*, 49(1), 71–5. doi: 10.1207/s15327752jpa4901_13.

Diener, E., Inglehart, R., and Tay, L. (2013). Theory and validity of life satisfaction scales. *Social Indicators Research*, 112(3), 497–527. doi: 10.1007/s11205-012-0076-y.

Emmons, R. A., and Diener, E. (1985). Personality correlates of subjective well-being. *Personality and Social Psychology Bulletin*, 11(1), 89–97. doi: 10.1177/0146167285111008.

Evans, G., and Lepore, S. J. (1997). Moderating and mediating processes in environment-behavior research. In *Advances in Environment, Behavior and Design*, ed. R. Moore and G. T. Marans, 255–285. New York: Plenum Press.

Fyhri, A., and Aasvang, G. M. (2010). Noise, sleep and poor health: Modeling the relationship between road traffic noise and cardiovascular problems. *Science of the Total Environment*, 408(21), 4935–4942. doi: 10.1016/j.scitotenv.2010.06.057.

Fyhri, A., and Klæboe, R. (2009). Road traffic noise, sensitivity, annoyance and self-reported health—A structural equation model exercise. *Environment International*, 35(1), 91–97. doi: 10.1016/j.envint.2008.08.006.

Gidlöf-Gunnarsson, A., and Öhrström, E. (2007). Noise and well-being in urban residential environments: The potential role of perceived availability to nearby green areas. *Landscape and Urban Planning*, 83(2–3), 115–126. doi: 10.1016/j.landurbplan.2007.03.003.

Gidlöf-Gunnarsson, A., and Öhrström, E. (2010). Attractive "quiet" courtyards: A potential modifier of urban residents' responses to road traffic noise? *International Journal of Environmental Research and Public Health*, 7(9), 3359–3375. doi: 10.3390/ijerph7093359.

Gidlöf-Gunnarsson, A., Öhrström, E., and Kihlman, T. (2010). A full scale intervention example of the quiet side concept in a residential area exposed to road traffic noise: Effects on the perceived sound environment and general noise annoyance. Presented at Internoise. Proc. of Internoise 2010, Lissabon, Sociedade Portuguesa de Acustica, Portugal.

Greenland, S. (2001). Ecologic versus individual-level sources of bias in ecologic estimates of contextual health effects. *International Journal of Epidemiology*, 30, 1343–1350. doi: 10.1093/ije/30.6.1343.

Guite, H. F., Clark, C., and Ackrill, G. (2006). The impact of the physical and urban environment on mental well-being. *Public Health*, 120, 1117–1126.

Hartig, T. (2004). Restorative environments. In *Encyclopaedia of Applied Psychology*, ed. C. Spielberger, 273–279. San Diego: Academic Press.

Hartig, T., Johansson, G., and Kylin, C. (2003). Residence in the social ecology of stress and restoration. *Journal of Social Issues*, 59(3), 611–636.

Hartig, T., Mitchell, R., de Vries, S., and Frumkin, H. (2014). Nature and health. *Annual Review of Public Health*, 35(1), 207–228. doi: 10.1146/annurev-publhealth-032013-182443.

Health Council of the Netherlands (HCN). (2006). Quiet areas and health. Publication 2006/12. The Hague: HCN.

Heinonen-Guzejev, M., Vuorinen, H. S., Mussalo-Rauhamaa, H., Heikkilä, K., Koskenvuo, M., and Kaprio, J. (2005). Genetic component of noise sensitivity: Twin research on human genetics. *Twin Research and Human Genetics*, 8, 245–249.

Houthuijs, D. J. M., and van Wiechen, C. M. A. G., eds. (2006). Monitoring van gezondheid en beleving rondom de luchthaven. Schiphol RIVM Rapport 630100003/2006.

Hume, K., and Ahtamad, M. (2013). Physiological responses to and subjective estimates of soundscape elements. *Applied Acoustics*, 74(2), 275–281. doi: 10.1016/j.apacoust.2011.10.009.

Hur, M., Nasar, J. L., and Chun, B. (2010). Neighborhood satisfaction, physical and perceived naturalness and openness. *Journal of Environmental Psychology*, 30(1), 52–59. doi: 10.1016/j.jenvp.2009.05.005.

ISO/TS 15666. (2003). Acoustics—Assessment of noise annoyance by means of social and socio acoustic surveys.

Jennings, P., and Cain, R. (2013). A framework for improving urban soundscapes. *Applied Acoustics*, 74(2), 293–299. doi: 10.1016/j.apacoust.2011.12.003.

Job, R. (1993). The role of psychological factors in community reaction to noise. In *Proceedings of the 6th International Congress on Noise as a Public Health Problem*, ed. M. Vallet, 48–59. Vol. 3. Arcueil, France: INRETS.

Job, R. (1996). The influence of subjective reactions to noise on health effects of the noise. *Environment International*, 22(1), 93–104. doi: 10.1016/0160-4120(95)00107-7.

Job, R. F. S. (1999). Noise sensitivity as a factor influencing human reactions to noise. *Noise and Health*, 3, 57–68.

Job, R. F. S., and Hatfield, J. (2001). The impact of soundscape, enviroscape, and psychscape on reaction to noise: Implications for evaluation and regulation of noise effects. *Noise Control Engineering Journal*, 49(3), 120–124. doi: 10.3397/1.2839647.

Job, R. F. S., Hatfield, J., Carter, N. L., Peploe, P., Taylor, R., and Morrell, S. (2001). General scales of community reaction to noise (dissatisfaction and perceived affectedness) are more reliable than scales of annoyance. *Journal of the Acoustical Society of America*, 110(0001-4966 SB-IM), 939–946.

Jones, D. M., Chapman, A. J., and Auburn, T. C. (1981). Noise in the environment: A social perspective. *Journal of Environmental Psychology*, 1(1), 43–59. doi: 10.1016/S0272-4944(81)80017-5.

Jylhä, M. (2009). What is self-rated health and why does it predict mortality? Towards a unified conceptual model. *Social Science and Medicine*, 69(3), 307–316. doi: 10.1016/j.socscimed.2009.05.013.

Kang, J., Chourmouziadou, K., Sakantamis, K., Wang, B., and Hao, Y., eds. (2013). *COST Action TD0804: Soundscape of European Cities and Landscapes*, 371. Oxford: COST.

Kaplan, R. (2001). The nature of the view from home: Psychological benefits. *Environment and Behavior*, 33(4), 507–542. doi: 10.1177/00139160121973115.

Kaptein, A. A., Helder, D. I., Kleijn, W. C., Rief, W., Moss-Morris, R., and Petrie, K. J. (2005). Modern health worries in medical students. *Journal of Psychosomatic Research*, 58, 453–457. http://www.uibno/ipq/pdf/B-IPQ-Dutch.pdf.

Kelfkens, G., Baliatsas, C., Bolte, J., and van Kamp, I. (2012). ECOLOG based estimation of exposure to mobile phone base stations in the Netherlands. In *Proceedings of the 7th International Workshop on Biological Effects of EMF*. Valletta, Malta: Electromagnetic Research Group (EMRG).

King, E. A., Murphy, E., and Rice, H. J. (2011). Implementation of the EU environmental noise directive: Lessons from the first phase of strategic noise mapping and action planning in Ireland. *Journal of Environmental Management*, 92(3), 756–64. doi: 10.1016/j.jenvman.2010.10.034.

Klæboe, R. (2005). Are adverse impacts of neighbourhood noise areas the flip side of quiet areas? *Applied Acoustics*, 68(5), 557–575.

Kroesen, M., Molin, E. J. E., Miedema, H. M. E., Vos, H., Janssen, S. A., and van Wee, B. (2010). Estimation of the effects of aircraft noise on residential satisfaction. *Transportation Research Part D: Transport and Environment*, 15(3), 144–153. doi: 10.1016/j.trd.2009.12.005.

Kroesen, M., Molin, E. J., and van Wee, B. (2013). Measuring subjective response to aircraft noise: The effects of survey context. *Journal of the Acoustical Society of America*, 133(1), 238–246.

Lachowycz, K., and Jones, A. P. (2013). Towards a better understanding of the relationship between greenspace and health: Development of a theoretical framework. *Landscape and Urban Planning*, 118, 62–69. doi: 10.1016/j.landurbplan.2012.10.012.

Lercher, P. (1996). Environmental noise and health: An integrated research perspective. *Environment International*, 22(1), 117–129. doi: 10.1016/0160-4120(95)00109-3.

Lercher, P., Evans, G. W., Meis, M., and Kofler, W. W. (2002). Ambient neighbourhood noise and children's mental health. *Occupational and Environmental Medicine*, 59(6), 380–6. doi:10.1136.

Lercher, P. (2007). Environmental noise: A contextual public health perspective. In *Noise and Its Effects*, ed. D. Luxon and L. M. Prasher, 345–377. London: Wiley.

Lercher, P. (2013). Health related quality of life and Environmental QoL in soundscape research and implementation. COST TUD0804, Final meeting, Merano/Meran, Italy.

Linder, S. H., and Sexton, K. (2011). Conceptual models for cumulative risk assessment. *American Journal of Public Health*, 101(Suppl. S1), S74–S81. doi: 10.2105/AJPH.2011.300318.

Maas, J., Verheij, R. A., Groenewegen, P. P., de Vries, S., and Spreeuwenberg, P. (2006). Green space, urbanity, and health: How strong is the relation? *Journal of Epidemiology and Community Health*, 60(7), 587–592.

Marans, R. W., and Stimson, R. J., eds. (2011). *Investigating Quality of Urban Life*, 453. Vol. 45. Dordrecht: Springer. doi: 10.1007/978-94-007-1742-8.

Martikainen, P., Laaksonen, M., Piha, K., and Lallukka, T. (2007). Does survey non-response bias the association between occupational social class and health? *Scandinavian Journal of Public Health*, 35, 212–215. doi: 10.1080/14034940600996563.

Meijman, T. F. (1991). Over vermoeidheid: Arbeidspsychologische studies naar de belevingvan belastingseffecten. Doctoral dissertation, University of Amsterdam.

Meijman, T. F., de Vries-Griever, A., de Vries, G. M., and Kampman, R. (1985). *The Construction and Evaluation of a One-Dimensional Scale Measuring Subjective Sleep Quality*. Groningen: Rijksuniversiteit.

Miilunpalo, S., Vuori, I., Oja, P., Pasanen, M., and Urponen, H. (1997). Self-rated health status as a health measure: The predictive value of self-reported health status on the use of physician services and on mortality in the working-age population. *Journal of Clinical Epidemiology*, 50(5), 517–528.

Mitchell, R., Astell-Burt, T., and Richardson, E. A. (2011). A comparison of green space indicators for epidemiological research. *Journal of Epidemiology and Community Health*, 65(10), 853–858.

Moser, G. (2009). Quality of life and sustainability: Toward person–environment congruity. *Journal of Environmental Psychology*, 29(3), 351–357. doi: 10.1016/j.jenvp.2009.02.002.

Öhrström, E., Skanberg, A., Svensson, H., and Gidlof-Gunnarsson, A. (2006). Effects of road traffic noise and the benefit of access to quietness. *Journal of Sound and Vibration*, 295(1–2), 40–59. doi: 10.1016/j.jsv.2005.11.034.

Ottosson, J., and Grahn, P. (2005). A comparison of leisure time spent in a garden with leisure time spent indoors: On measures of restoration in residents in geriatric care. *Landscape Research*, 30(1), 23–55.

Pacione, M. (2003). Urban environmental quality and human wellbeing: A social geographical perspective. *Landscape and Urban Planning*, 65(1–2), 19–30. doi: 10.1016/S0169-2046(02)00234-7.

Payne, S. R. (2013). The production of a Perceived Restorativeness Soundscape Scale. *Applied Acoustics*, 74(2), 255–263. doi: 10.1016/j.apacoust.2011.11.005.

Petrie, K. J., Sivertsen, B., Hysing, M., Broadbent, E., Moss-Morris, R., Eriksen, H. R., and Ursin, H. (2001). Thoroughly modern worries: The relationship of worries about modernity to reported symptoms, health and medical care utilization. *Journal of Psychosomatic Research*, 51(1), 395–401.

Pijanowski, B. C., Villanueva-Rivera, L. J., Dumyahn, S. L., Farina, A., Krause, B. L., Napoletano, B. M., Gage, S. H., and Pieretti, N. (2011). Soundscape ecology: The science of sound in the landscape. *Bioscience*, 61(3), 203–216. doi: 10.1525/bio.2011.61.3.6.

Riedel, N., Scheiner, J., Müller, G., and Köckler, H. (2013). Assessing the relationship between objective and subjective indicators of residential exposure to road traffic noise in the context of environmental justice. *Journal of Environmental Planning and Management*, 57(9), 1398–1421. doi: 10.1080/09640568.2013.808610.

Rodiek, S. D. (2002). Influence of an outdoor garden on mood and stress in older persons. *Journal of Therapeutic Horticulture*, 13, 13–21.

Schreckenberg, D., Griefahn, B., and Meis, M. (2010). The associations between noise sensitivity, reported physical and mental health, perceived environmental quality, and noise annoyance. *Noise and Health*, 12, 7–16.

Schreurs, P. J. G., van de Willige, G., Brosschot, J. F., Tellegen, B., and Graus, G. M. H. (1993). *De Utrechtse Coping Lijst: ucl, omgaan met problemen en gebeurtenissen.* Amsterdam: Harcourt Assessment.

Shepherd, D., Welch, D., Dirks, K. N., and Mathews, R. (2010). Exploring the relationship health-related quality of life in a sample of adults exposed to environmental noise. *International Journal of Environmental Research and Public Health*, 7, 3579–3594.

Sirgy, M. J. (2012). *The Psychology of Quality of Life*, 379. Vol. 50. Dordrecht: Springer. doi: 10.1007/978-94-007-4405-9.

Skånberg, A., and Öhrström, E. (2002). Adverse health effects in relation to urban residential soundscapes. *Journal of Sound and Vibration*, 250(1), 151–155. doi: 10.1006/jsvi.2001.3894.

Smith, J. W., and Pijanowski, B. C. (2014). Human and policy dimensions of soundscape ecology. *Global Environmental Change*, 28, 63–74. doi: 10.1016/j.gloenvcha.2014.05.007.

Søgaard, A. J., Selmer, R., Bjertness, E., and Thelle, D. (2004). The Oslo Health Study: The impact of self-selection in a large, population-based survey. *International Journal for Equity in Health*, 3(1), 3. doi: 10.1186/1475-9276-3-3.

Stansfeld, S. A., Clark, C. R., Jenkins, L. M., Tarnopolsky, A. (1985). Sensitivity to noise in a community sample: Measurement of psychiatry disorder and personality. *Psychological Medicine*, 15, 243–254.

Taylor, S. E., Repetti, R. L., and Seeman, T. (1997). Health psychology: What is an unhealthy environment and how does it get under the skin? *Annual Review of Psychology*, 48, 411–47. doi: 10.1146/annurev.psych.48.1.411.

Terluin, B. (1996). De Vierdimensionale Klachtenlijst (4DKL). Een vragenlijst voor het meten van distress, depressie, angst en somatisatie. *Huisarts Wet 1996*, 39, 538–547.

van der Zee, K. I., and Sanderman, R. (1993). RAND-36. Groningen: Northern Centre for Health Care Research, University of Groningen, the Netherlands, ISBN 90721560928.

van Kamp, I. (1990). *Coping with Noise and Its Health Consequences*, 2009. Groningen: STYX and PP Publications.

van Kamp, I., and Davies, H. (2013). Noise and health in vulnerable groups: A review. *Noise and Health*, 15(64), 153–159.

van Kamp, I., Leidelmeijer, K., Marsman, G., and de Hollander, A. (2003). Urban environmental quality and human well-being. *Landscape and Urban Planning*, 65(1–2), 5–18. doi: 10.1016/S0169-2046(02)00232-3.

Van Kempen, E., van Kamp, I., and Kruize, H. (2011). The need for and access to quiet areas in relation to annoyance, health and noise-sensitivity. In *10th International Congress on Noise as a Public Health Problem (ICBEN) 2011*, London, p. 441.

Van Renterghem, T., Botteldooren, D., and Lercher, P. (2007). Comparison of measurements and predictions of sound propagation in a valley-slope configuration in an inhomogeneous atmosphere. *Journal of the Acoustical Society of America*, 121(5), 2522–2533.

Voigtländer, S., Vogt, V., Mielck, A., and Razum, O. (2014). Explanatory models concerning the effects of small-area characteristics on individual health. *International Journal of Public Health*, 59(3), 427–438. doi: 10.1007/s00038-014-0556-8.

Wallenius, M. A. (2004). The interaction of noise stress and personal project stress on subjective health. *Journal of Environmental Psychology*, 24, 167–177.

Weden, M. M., Carpiano, R. M., and Robert, S. A. (2008). Subjective and objective neighborhood characteristics and adult health. *Social Science and Medicine*, 66(6), 1256–70. doi: 10.1016/j.socscimed.2007.11.041.

White, K., Hofman, W. F., and van Kamp, I. (2010). Noise sensitivity in relation to baseline arousal, physiological response and psychological features to noise exposure during task performance. Proceedings of Internoise 2010, Lissabon, Sociedade Portuguesa de Acustica, Portugal.

World Health Organization. (2013). *Health 2020: A European Policy Framework and Strategy for the 21st Century*, 182. Copenhagen: WHO Regional Office for Europe.

Yzermans, J., Baliatsas, C., van Dulmen, S., and van Kamp, I. (2012). Assessing non-specific symptoms in epidemiological studies: Development and validation of the Symptoms and Perceptions (SaP) questionnaire. *International Journal of Hygiene and Environmental Health*. Available online 31 August 2015.

Chapter 6

Human Hearing–Related Measurement and Analysis of Acoustic Environments

Requisite for Soundscape Investigations

Klaus Genuit[1,2] and André Fiebig[1,2]
[1]HEAD acoustics, Herzogenrath, Germany
[2]Eberstr 30a 52134, Herzogenrath, Germany

CONTENTS

Binaural hearing allows for localizing sound sources, suppressing noise, for example, for a better understanding of speech, and forming an impression of spaciousness. In order to analyze acoustical environments appropriately, it is necessary to use human hearing–related measurement methods. For example, measurements with binaural technology enable us to reexperience the acoustic environment in an acoustically sense. Moreover, by means of the knowledge of psychoacoustics, several basic auditory sensations can be described in detail beyond the impression of loudness. This chapter introduces binaural measurement technology and psychoacoustics, which are considered necessary components of a comprehensive soundscape investigation.

6.1 MEASUREMENT EQUIPMENT

6.1.1 Introduction to Anatomy and Signal Processing of Human Hearing

Human hearing works in many respects differently than conventional sound measuring systems. In contrast to an omnidirectional microphone, the human ear is a directional filter, which leads to sound pressure level (SPL) at the ear drum varying by +15 to −30 dB, related to the unfiltered signal depending on frequency and direction of sound incidence. The filtering properties are caused by diffractions and reflections, depending on direction, as well as by resonances, which are independent of direction. The noise changes are caused by head, shoulder, torso, and pinna. In contrast to it, a standard measuring microphone has a linear, frequency-independent response characteristic for all directions of sound incidence. Figure 6.1 illustrates the specific pattern, which is created due to the directional filtering.

The most obvious difference between human hearing and an omnidirectional microphone is that humans use two auditory paths with mostly different input signals, which permit binaural signal processing, source localization, pattern recognition, and distance localization to a certain degree. Human hearing comprises these two input channels, which, together with selectivity, create spatial hearing and therefore, in the case of several spatially distributed sound sources, yield results different from those provided by monaural measuring procedures. The localization of sound sources is possible in both the horizontal and the median plane. In the horizontal

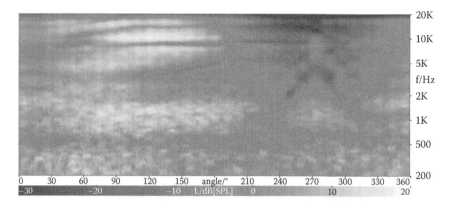

Figure 6.1 Directional filter characteristics of human hearing. A source emitting a white noise is moving around an artificial head (360°). The changed noise due to the directional filtering of one ear compared to the reference signal is shown.

plane, the localization is based on the evaluation of interaural differences. As soon as the sound source is not located exactly in front or back of the listener, the different distances to the two ears cause interaural delays. The human brain is able to interpret these delays (less than 1 millisecond) as directional information. These delays reach their maximum if the sound source is located to the left or right of the listener. The interaural time differences (ITDs) are considered the most important basis for the localization of sound sources. In addition, interaural level differences (ILDs) are caused by acoustic shadowing. Such level differences are also interpreted by humans and are used for localization. Both effects are applied for spatial orientation. The localization of sound sources in the median plane is based on a different phenomenon. If a sound source is moving along the median plane, only monaural cues exist (Vorländer, 2008). Yet humans are capable of localizing sound sources in the median plane as well. This capability is based on the direction-dependent filtering of sound caused by the anatomic shape of the auricles, head, shoulders, and upper human body. Depending on the direction of sound incidence, the frequency spectrum of the sound reaching the ears is changed. These spectral differences can be interpreted as directional information, and certain distortion patterns are associated with specific directions (Blauert, 1996). Compared to the localization in the horizontal plane, the localization in the median plane is less accurate.

Thus, the directional filtering of the human ear causes specific hearing phenomena and allows for spatial hearing, selectivity, and binaural noise suppression. Exemplarily, Figure 6.2 illustrates the need for binaural measurement equipment. Two spatially distributed loudspeakers emit two different signals: a pulsed sine with a frequency of 4 kHz (left loudspeaker) and a narrowband noise with a centre frequency of 4 kHz and a bandwidth of 760 Hz (right loudspeaker). It can be clearly seen that in case of two omnidirectional microphones, which were positioned at the same distance as the microphones of an artificial head, the pulsed tone is fully masked, even in case of the microphone closer to the pulsed sine source (left microphone). In contrast to it, due to the filtering properties of the human hearing, the tone is not masked in case of the left artificial head channel, which faces more the loudspeaker with the pulsed 4 kHz sine. It is clear that pre-, post-, and simultaneous masking properties are different from a binaural point of view, if the masking and the masked signal have different directions. Humans exploit these possibilities and, for example, enhance the speech intelligibility in noisy environments by means of binaural signal processing.

Moreover, humans have a high resolution in the frequency and time domain in combination with a high dynamic range of more than 120 dB. For ordinary analyzers, the product of spectral resolution multiplied with temporal resolution equals 1. From psychoacoustic investigations, it is known that this product amounts to 0.3 for humans, which means that humans use a high-frequency resolution and, at the same time, can perceive fast temporal variations, such as short amplitude or frequency changes (Genuit, 1992b).

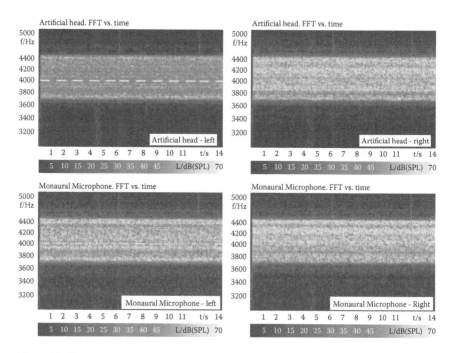

Figure 6.2 Two loudspeakers emitting noise from different directions (−60° and +60°). The left loudspeaker played back a pulsed sine with a frequency of 4 kHz, and the right loudspeaker played back a narrowband noise with a centre frequency of 4 kHz. Top: Artificial head recording (left and right channels). Bottom: Recording of two omnidirectional microphones, which were positioned at a distance comparable to that of the microphones of the artificial head.

These statements illustrate the extraordinary high performance of human hearing with respect to the processing of sound, which is very difficult or still impossible to match by current technical devices and analyses.

6.1.2 Derivation of Requirements for Measurement Equipment

On the basis of conventional one-channel microphone measurements, it is not possible to sufficiently grasp the physical entity of an acoustic environment and consider the human way of hearing adequately. Because of complex environmental sound situations caused by several spatially distributed natural and technical sound sources, it is important to use binaural technology in order to consider masking effects, spatial distribution, and complex phase relations. This means that the influence of the (artificial) head with its transfer functions on the respective sound field should not be interpreted as an error, but represents the way humans perceive sound. For example, listening experiments have shown that physiological reactions to moving and spatially distributed sound sources are stronger than, for example, the reaction to noise caused by motionless sources or unidirectional noise. It was found that the aurally accurate playback of real industrial noises caused by spatially distributed sources recorded by an artificial head causes stronger psychological reactions than the same industrial noise situation recorded with a conventional monaural microphone (Genuit et al., 1997).

Measurements with binaural technology are of particular importance if subsequent (aurally accurate) reproductions of noise are required, for example, in the case of further examination of the sounds in laboratory listening tests (Genuit and Fiebig, 2006). "Copies" of the acoustic environment as close as possible to human perception are needed, especially regarding archiving and reexperiencing the acoustic scenery for comparability and analysis reasons.

Thus, aurally accurate sound measuring technology is not an alternative to, but an important extension of existing sound measurement techniques. In complex sound situations, this technology gathers additional data necessary for the comprehensive, aurally equivalent evaluation of sound.

6.1.3 Introduction to Binaural Measurement Equipment

Binaural technology, the key component of aurally equivalent measurements, comprises recording of sound by means of an artificial head measuring system.

The principal idea of aurally accurate recordings and playback is shown in Figure 6.3. Two microphone signals are recorded and equalized to achieve signals compatible with conventional monaural microphone output.

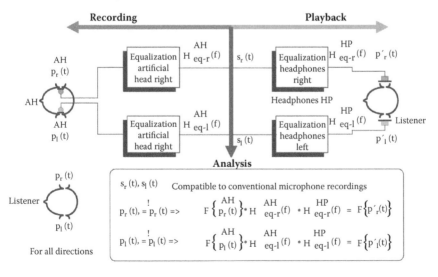

Figure 6.3 Principle of head-related transmission.

For playback, the respective signals are equalized by correction filters to create the same signals in the ear canal of the listener as if the listener had been in the original sound situation.

It is necessary to pay special attention to the correct positioning of all acoustically relevant parts of the body—in particular the position of the artificial auricles relative to the head and shoulder parts. The angles of inclination and aperture also have a significant influence on the head-related transfer functions. Positioning errors become apparent in the direction-dependent part of the transfer function and cannot be corrected after the measurement.

Today, artificial head measurement systems are available that have realized transfer functions comparable to those of human hearing with a high dynamic range. Since it is well known how the geometry of a head influences recorded signals, artificial heads could be simplified without losing an average directional pattern. Thus, it was possible to develop artificial heads that correspond to a mathematically describable geometry with an average directional pattern representing that of human hearing. The dimensions comply with international standards; the free-field transfer function and the directional pattern are in accordance with the (IEC 959, 1990) report.

Figure 6.4 shows the transfer functions of an artificial head measurement system in the horizontal plane. The transfer functions are measured monaurally at the left ear, and the sound pressure levels are always related to sound pressure level at frontal sound incidence (0°). Figure 6.5 displays an example of a commercial head measurement system, which possesses a fully mathematically describable geometry.

In addition to the well-known free-field and diffuse-field equalization, an independent direction equalization (ID equalization) was introduced, which compensates only the direction-dependent part of the artificial head transfer function (Genuit, 1992a). The direction-independent components are caused by resonances at the cavum conchae and the ear canal entrance.

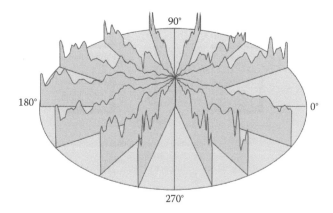

Figure 6.4 Monaurally measured transfer functions (left ear) related to sound pressure level at frontal sound incidence (0°).

Figure 6.5 Artificial head measurement system with mathematically describable simplified geometry.

In particular, in the case of an artificial head based on mathematically describable, simplified geometry, the direction-independent components can be determined precisely for the purpose of equalization.

6.2 MEASUREMENT CONDITIONS AND SPECIFICATIONS

6.2.1 Derivation of Measurement Specifications due to Human Hearing Performance

Sound recordings should be digitally stored as time signals. Since the audible frequency range of human beings goes approximately from 16 Hz to 16 kHz, the sampling rate should not go below 44.1 kHz due to the known Nyquist–Shannon sampling theorem. The dynamic range should cover the sound-pressure-level range, which is relevant for human listeners. Humans have a dynamic range up to 120 dB. Thus, an artificial head should operate with a dynamic range comparable to that to avoid clipping of audible noises. Commercial artificial head systems provide dynamic ranges (S/N) of more than 100 dB.

Figure 6.6 Artificial head measurement system with windshield to avoid disturbing wind turbulences.

Since wind turbulences can greatly disturb noise measurements, the use of windshields, in particular for outdoor measurements, is imperative (see example in Figure 6.6). A windshield blocks disturbing wind noise in outdoor recordings. Its influence on the frequency response for recorded signals is very low. It is also possible to consider suitable windshield equalization to correct the effect caused by the device.

Moreover, to reduce the risk of overload caused by irrelevant noises, the application of an analogue high-pass filter is recommended. Low-frequency interferences, such as opening and closing of doors, can be suppressed by activating high-pass filters. High-pass filters are used to attenuate undesired background noise that could overmodulate the A/D converter. This allows a higher sensitivity setting to be used, which leads to a higher dynamic range for sounds in the relevant audible frequency areas. Usually, different high-pass filters are available, such as a third-order high-pass filter with a cutoff frequency of 22 Hz or higher.

6.2.2 Derivation of Measurement Conditions

The soundscape approach has frequently shown that it is necessary to implement new methods and approaches in order to fully understand the human perception of noise in a specific environment and ambiance. Soundscape researchers are interested in the answer to the question of not only *how much*, but also *why*. Only then is a reliable interpretation of the *how much* dimension possible (Genuit and Fiebig, 2014).

According to ISO 1996-2, the height of microphones should be chosen according to the actual or expected height of the receiver (ISO 1996-2). Since in soundscape investigations usually typical receiver positions are considered as well, this definition appears to be adequate for soundscape investigations. Of course, the measurement height cannot follow the principle of noise maps, where the sound-pressure-level calculations are related to a height of 4 m. It is reasonable to apply a typical height of adult receiver ranging from 1.5 to 2 m.

In contrast to the "indoor" noise measurement position according to ISO 1996-2, where the microphone position is defined to be within 0.5 m of the front of an open window, this receiver position appears inadequate in a soundscape study. In a soundscape investigation, the acoustical measurement must consider only relevant receiver positions. This means that comparable to the American standard (ANSI S12.9), where it is defined that the choice of the measurement positions depends on the *purpose of the measurements*, the measurement spots in soundscape investigations must consider the respective type of investigated soundscape, which provides the relevant receiver positions.

The selection of an adequate measurement time interval is also a vital boundary condition in a soundscape investigation to have reliable data for analysis. The ISO 1996 standard defines that the reference time intervals shall be chosen to take into account typical human activities and variations in the operation of the noise source (ISO 1996-3). With respect to soundscape investigations, the term *variations in the operation of the noise source* should be adapted to *variations caused by the prominent sound sources*. Prominent sound sources represent sources that could be classified with soundscape-related terms such as signals, sound marks, or keynote sounds (Schafer, 1977). It is very important to emphasize that prominence does not mean highest sound pressure levels. The sound sources are prominent, which attract attention due to its meaning, specific acoustical properties, or visual peculiarity (for more information, see Chapter 2).

In the European research project IMAGINE (2006), it is defined that the duration of a noise measurement should be sufficiently long to encompass all emission situations that are needed to obtain a representative average. With respect to soundscape investigations, this definition should be supplemented with the expression to obtain a representative *picture* of the whole soundscape, with its expected important and typical sounds.

However, it goes without saying that several elements and components of the soundscape concept cannot be simply integrated into existing environmental noise standards and guidelines. First, the application of binaural measurement technology for measuring acoustical environments is not defined in any environmental noise standards, and its use is not supported in any current legal framework so far. However, it was described above that binaural measurements are imperative, since it is necessary to provide a good sense of spatial immersion in the recreated acoustic environments to ensure ecological validity (Dubois et al., 2006).

Moreover, for example, ISO 1996 proposes that measurement positions should be chosen in the vicinity of each of the sources in order to reduce the influence of the other sources. This does not correspond to the soundscape concept. In a soundscape investigation, the main focus lies not in the separation of the different source contributions, but in recording and analyzing *acoustical environments as a combination of all relevant sound sources.* The separation of the contributions of the different sound sources is considered for analytical and legal reasons in noise policy, but in soundscape study, the examination of the whole remains inevitable (Fiebig and Genuit, 2011). The concept of the soundscape as a "musical composition" (Schafer, 1977) requires us to analyze the acoustic environment as a whole, as well as in its different facets, comparable to the study of music.

Another aspect concerns the preferred noise descriptor. In accordance with ISO 1996-3, the preferred noise descriptor for the specification of noise limits is the equivalent continuous A-weighted sound pressure level (L_{Aeq}). Although there is no agreement upon the most valuable acoustic indicator regarding the description of the acoustic environment in the scope of soundscapes so far, it is very likely that the simple time-averaged sound pressure level, such as the L_{Aeq}, will not turn out to be the preferred and most important noise indicator. This issue will be discussed in the following chapters more in detail.

Since there is no established or standardized measurement procedure in the field of soundscape studies so far, it is very important to list the basic data with respect to equipment, measurement procedure, instrumentation, and so forth. It includes the exact positioning of the artificial head measurement system (height, orientation), as well as information about measurement time, duration, a detailed description of measurement points, the applied measurement equipment, and the atmospheric conditions (like weather, wind, temperature, humidity). This could be complemented by notations of influences of topographical features and local shielding effects or a description of the sound sources. Such a detailed measurement protocol is imperative for the analysis of comparability of soundscape studies.

6.3 ANALYSIS OF ACOUSTIC ENVIRONMENTS BY MEANS OF PSYCHOACOUSTICS

6.3.1 Introduction to Psychoacoustics

Psychoacoustics deals with the quantitative link between physical stimuli with their caused hearing sensations (Fastl and Zwicker, 2007). It describes sound perception mechanisms in terms of specific parameters, such as loudness, sharpness, roughness, and fluctuation strength. In general, the approach regarding the investigation of certain aspects of basic auditory sensations is to derive mathematical descriptions from measured relationships between stimulus and perception, which form the basis for a model, describing the manner of human noise perception (Sottek, 1993).

For example, the psychoacoustic parameter *loudness* was introduced several decades ago, which considers the basic human hearing processing phenomena with respect to the sensation of volume. This parameter considers basic human signal processing effects like spectral sensitivity (frequency weighting), masking (post- and simultaneous), critical bands, and non-linearities. In all, the loudness sensation is related to the spectral content of a sound and to the human signal processing.

Due to the used algorithms mimicking the human hearing loudness sensation process, the psychoacoustic parameter loudness offers advantages over the A-weighted sound pressure level. It was found that this parameter shows a much better correspondence with loudness sensation than the L_{Aeq} (Bray, 2007). It has to be emphasized that the psychoacoustic loudness goes far beyond simple sound-level indicators (Genuit, 2006). The advantage of loudness compared to A-weighted sound-pressure-level indicators becomes even clearer when the superposition of sounds is considered. For example, when two sounds with different spectral shapes are combined, the A-weighted SPL is unable to predict the perceived loudness. It becomes even more complicated when two tones are added to noise. In all, the A-weighted SPL can be inversely related to loudness and annoyance (Hellman and Zwicker, 1987).

Figure 6.7 shows the analysis of loudness of three simple noises with an identical A-weighted SPL. Since the psychoacoustic loudness can be interpreted as a ratio-scaled quantity, the values illustrate the great mismatch between the psychoacoustic loudness indicator and the A-weighted time-averaged SPL. The broadband noise is perceived as twice as loud as the narrowband noise, although the dB(A) indicator does not indicate any difference.

Another interesting phenomenon is illustrated by the police siren sound, which is strongly time varying. It has the same time-averaged A-weighted SPL as the synthetic sounds, but varies strongly over time. The representative single loudness value of 34.6 sone is only reached or exceeded in very few moments. This is due to the fact that the cognitive stimulus integration of humans is complex; a human does not simply average his or her sensational level over time (Stemplinger, 1999). The statistical mean of a time-variant loudness run leads to results that are too low in comparison to evaluated overall loudness. Thus, the percentile loudness N5 has to be determined when stating the perceived overall loudness (DIN 45631/A1). N5 means the loudness value, which is reached or exceeded only 5% of the total time.

Moreover, there is strong evidence that physiological reactions to noise correlate better with the loudness parameter than with the sound pressure level. Jansen and Rey (1962) had already shown several decades ago that the finger pulse amplitude after exposure to different sounds with the same sound pressure level can strongly vary. The variances can be explained on the basis of the differences in loudness (Genuit and Fiebig, 2007a).

The psychoacoustic parameter *sharpness* describes the perception of the spectral centre of a signal with emphasis on high frequencies. In short,

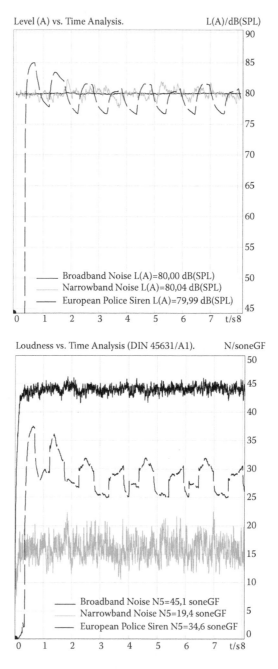

Figure 6.7 Comparison of sound pressure level and loudness analysis results of three different sounds: broadband noise (20–16,000 Hz), narrowband noise (950–1050 Hz), and police siren sound. Top: Level (A-weighted) vs. time. Bottom: Loudness according to the DIN 45631/A1 vs. time.

the relationship between the weighted specific loudness values with special emphasis on the high critical bands to the total loudness is determined. Sharpness is calculated from loudness and expresses the spectral balance of a sound; more high-frequency content raises the sharpness impression.

The calculation of the psychoacoustic parameter sharpness is defined in the German standard DIN 45692. Besides the German standard, other methods for sharpness calculation are available, such as Aures (1984) or Bismarck (1974). The DIN 45692 standard and Bismarck calculation method produce similar sharpness results. Unlike these methods, the sharpness according to the Aures calculation method rises in value for a constant spectral shape as loudness rises; Aures defined the sharpness impression to be dependent on the total loudness.

It is usually observed that sensory pleasantness decreases with increasing sharpness. Figure 6.8 shows that remarkable sharpness differences can occur in the context of environmental noises.

The psychoacoustic parameters *roughness* and *fluctuation strength* deal with perception of temporal effects. The roughness metric quantifies the

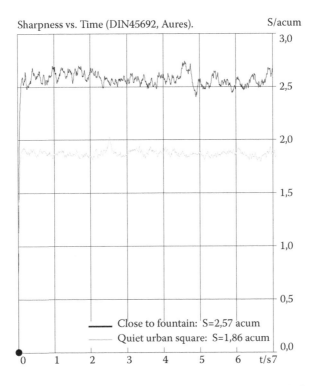

Figure 6.8 Comparison of sharpness of two environmental sounds. Black: Environmental noise recorded close to a fountain. Gray: Environmental noise recorded at a quiet public square. Sharpness according to the Aures calculation method over time.

perceived strength of rapid modulation between the rates of about 20 and 300 Hz. For constant sound pressure and modulation strength, the roughness for beating tones or modulated noises matches a specific hearing sensation, which peaks at a modulation rate of 70 Hz (Fastl and Zwicker, 2007). It has to be mentioned that several roughness calculation methods return exceptionally high values for broad- or narrowband noise with stochastic variations, which are not perceived as rough at all. Thus, the roughness metric must be used cautiously (Bray, 2007).

In contrast to roughness, the fluctuation strength parameter quantifies the perception of slow modulations between just perceptible rates and a rate of about 20 Hz with a maximum at 4 Hz modulation.

The psychoacoustic parameter *tonality* describes the sensation related to the amount of prominent tones or narrowband components in a signal. To identify prominent and salient tonal components, *prominence ratio* analysis can be applied (Nobile and Bienvenue, 1991). The prominence ratio determines power output in a band of critical bandwidth. This is then related to the average power output in both of the neighbouring critical bands. This means that, again, not the absolute level of an identified tonal component is most important, but rather its relationship to surrounding critical bands. Table 6.1 lists basic psychoacoustic parameters and their meanings.

A hearing-related parameter that has shown its significance in several surveys (Fiebig et al., 2009) is the relative approach parameter, which is related to the perceivable patterns in acoustic signals (Genuit, 1996). This analysis detects specific, obtrusive, attention-attracting noise features. It is often observed that human auditory impression is dominated by patterns and, in such situations, largely ignores absolute values (Genuit and Bray, 2006).

The relative approach analysis simulates the ability of the human hearing to adapt to stationary sounds and react, on the other hand, to variations and patterns in the time and frequency structure of a sound. The algorithm

Table 6.1 Basic Psychoacoustic Parameters and Their Meaning

Parameter	Description	Consideration of ...
Loudness	Perception of volume (loudness)	Critical bandwidths and masking properties of human hearing
Sharpness	Perception of the spectral centre of a signal with emphasis on high-frequency content	Relationship between the weighted specific loudness values with special emphasis on higher critical bands to the total loudness
Roughness	Perception of fast modulations between 20 and 300 Hz	Fast modulations (amplitude and frequency modulation)
Fluctuation strength	Perception of slow modulations below 20 Hz	Slow modulations (amplitude and frequency modulation)
Tonality	Perception of prominent tones in a signal	Amount of prominent tonal (or narrowband) components

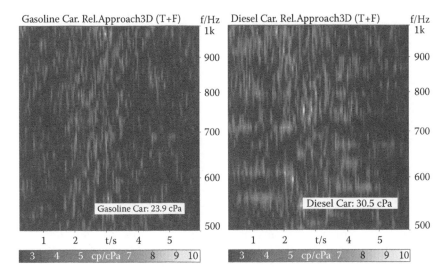

Figure 6.9 Comparison of pass-by events of gasoline (left) and diesel (right) engine vehicles. Relative approach analysis shows a greater amount of patterns in the diesel car pass-by event. Both events have the same A-weighted sound pressure level.

forms a moving average in time and frequency and subtracts it from each series of instantaneous measurements; the difference between instantaneous value and the updating moving average is interpreted as pattern strength. It is the difference between the estimate and the actual content in a signal.

In Figure 6.9, two vehicle pass-by events are analyzed that have the same sound pressure level (L_{Aeq}). It can be clearly seen that the relative approach analysis identifies the diesel knocking of the second pass-by event, which is perceived as more annoying by the majority of persons.

Another example of a relative approach analysis is shown in Figure 6.10. Two short sequences of environmental noises recorded in Aachen are analyzed with the acoustical pattern quantification analysis in the time and frequency domain. It can be seen that in case of the pseudostationary noise caused by the water splashing of a fountain, any remarkable noise patterns are not recognized, whereas in case of the carillon, a great strength of patterns can be found. These patterns dominate the perception.

It was already mentioned that human hearing quickly adapts to stationary signals, but remains very sensitive to fluctuations and intermittent noise, as well as to prominent, salient noise events. Therefore, peak values and relative changes can be significant with respect to the auditory perception as well. Percentile levels allow the consideration of fluctuations and variations over time (Genuit, 2006).

Figure 6.11 shows exemplarily a loudness run of an acoustical outdoor measurement performed in Pisa. Based on such a loudness run, percentile

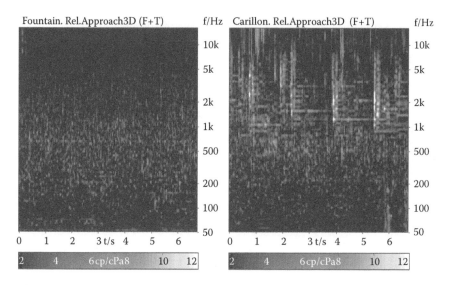

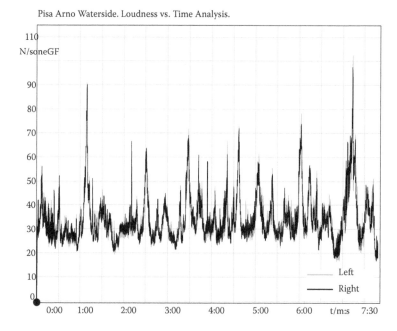

Figure 6.10 Comparison of two environmental noise measurements. Left: Noise recorded near a fountain. Right: Public space and the noise of a carillon recorded. Relative approach analysis clearly shows the pseudostationary character of the fountain noise and the great amount of acoustical patterns produced by the chimes.

Figure 6.11 Loudness analysis according to DIN 45631/A1 of an environmental noise measured in Pisa Arno Waterside.

values can be calculated, such as N90 describing the background loudness or N5 loudness, the value that is reached or exceeded in only 5% of the measurement time. It can be recognized that in the considered example, prominent loud scooters occur, but several times the loudness drops to a low value (periods of quietness). The peak loudness events are mainly caused by motor scooters (Genuit and Fiebig, 2007b). To grasp such fluctuations and variations, the calculation of relative percentile value differences is helpful (e.g., ×5 to ×50 or ×5 to ×90). In all, Pisa has strong loudness variations, which can be described by the ratio of N5 to N90. Here, the relative difference between N5 and the N90 indicator is about 60%, which illustrates the strong variations in the loudness run.

In general, the difference between the high and low loudness percentile values is an indicator for environmental noise quality (Genuit, 2006). Great differences indicate a strong unsteadiness with respect to loudness. Such loudness variations usually attract more attention than less varying noise.

6.3.2 Psychoacoustic Analysis of Acoustic Environments (Soundwalk Data) and Basic Classification of Acoustic Environments (Soundscapes) Based on Psychoacoustic Profiles

Although the need for advanced acoustic indicators mimicking the way humans perceive and process sound is almost undisputed, environmental noise assessment and control are still exclusively based on laws, guidelines, and regulations, which refer only to simple (typically long-term-averaged) sound-pressure-level indicators. Although it is widely accepted that the measurement and interpretation of sound pressure levels failed in adequately describing environmental noise annoyance, the noise policy is still driven by very simple assumptions. These simple indicators are well established and deeply ingrained in noise policy over decades. In contrast, psychoacoustic calculation methods, as functions of time structure, sound pressure level, and spectral distribution, allow a more detailed description of environmental noises.

Several case studies investigating the soundscapes of Aachen were performed by researchers participating in training schools in the context of the COST Action TD 0804. Three events took place, where more than 40 young researchers coming from all over Europe have studied the soundscapes of Aachen.

It is clear that with small group sizes, statistically significant results cannot be derived. However, the soundwalks repeated over 3 years allow an insight into the acoustical robustness of acoustical environments and their interindividual impacts on visitors. The soundwalks were carried out three times by two groups of participants examining eight defined locations in Aachen city by walking in opposite directions; group 1 started at place 1, and at the same time group 2 walked from place 8 to place 1 (Figure 6.12). The

Figure 6.12 Investigated locations during soundwalks through the city of Aachen.

eight sites with different characteristics were chosen along a route starting from a historical gate, to Aachen city centre, ending at a historical fountain (Fiebig et al., 2010) (see Figure 6.12). At each location, the sound was recorded with a mobile front end and a binaural headset, and the impressions as well as assessments were written down by participants on an evaluation sheet. At each location, each of the two groups carried out recordings with measurement duration of 3 minutes. The measurements were done with fixed orientation of the headset and a fixed position. Simultaneously, one participant took photos of the location, and all participants have used all senses for their evaluations (free words as well as unpleasantness and loudness ratings on a 5-point continuous scale ranging from "not at all" to "extremely").

Figure 6.13 displays exemplarily the similarity in the spectra of the recordings of the same acoustical environment measured over 3 years. The road is only open for public transportation, which causes only intermittent road traffic events of heavy vehicles (buses), leading to obtrusive single broadband noise events. Buses emit low-frequency engine noise below 200 Hz. Moreover, high-frequency noise caused by small fountains at ground level is observable in all spectra as well.

Exemplarily, the psychoacoustic loudness of measurement locations 01 and 08 was analyzed over diverse measurements with a length of 3 minutes. Figure 6.14 shows different loudness percentile values (according to DIN 45631/A1, N5 loudness represents the perceived average loudness of time-variant noises; the N90 loudness indicator represents a kind of background loudness). It can be seen that although the loudness of the background noise is almost similar, location 1 has higher loudness peaks expressed by N5 values. The only exception occurred in 2011, where during the 3-minute recording of group 1 an ambulance car passed by, leading to a high N5 loudness value. The lower diagram in Figure 6.13 shows the ratio

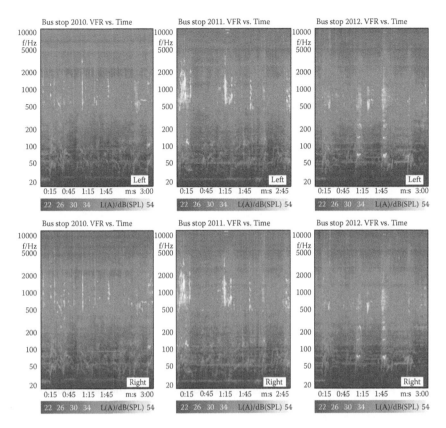

Figure 6.13 Comparison of the noise at location 08 measured over 3 years (left, 2010; middle, 2011; right, 2012; top, left channel; bottom, right channel). Variable frequency resolution (VFR) spectrum vs. time.

between the loudness peaks (N5) and the loudness of the background noise (N90) as an indicator for loudness variation. A small ratio value stands for a low variation of loudness over time, whereas a high ratio value indicates large variations of loudness over time. This ratio value is helpful in understanding human responses to noise, which are related not only to absolute values, but also to relative values and patterns. As expected, the judgements given in situ correspond to the loudness characterization of the two locations. Forty-two subjects rated location 08 on a continuous 5-point scale as significantly less annoying (3.42–4.02) and less loud (3.14–3.76).

Although the recordings of the different locations made by two groups were time-shifted in each year, the 3-minute fragments of the investigated soundscapes showed comparable psychoacoustic properties and patterns, as exemplarily shown in Figure 6.13. To investigate similarities of the different recordings in principle, the recordings made over 3 years were subject to a cluster analysis. Input variables were only psychoacoustic properties of

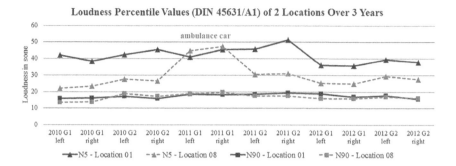

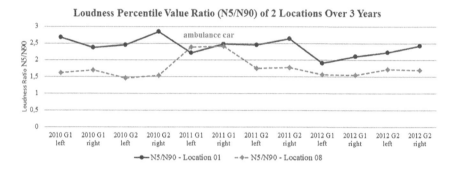

Figure 6.14 Loudness analyses of measurements performed at locations 01 and 08. Top: N5 and N90 values of both locations over the years and groups. Bottom: Loudness parameter ratio N5/N90 of both locations over the years and groups.

the measured soundscapes, such as loudness, sharpness, roughness, impulsiveness, and Relative Approach values. Each soundscape (measurement point) was considered three times (2010, 2011, 2012) in the cluster analysis, since each year measurements were performed. The psychoacoustic values calculated for both measurements per year were averaged (mean). Thus, the 48 recordings were merged to 24; for each measurement place and each year, one object was created. Figure 6.15 presents the result of the cluster analysis (average linkage method, Euclidean distance).

The cluster analysis result illustrates that based on the psychoacoustic properties of recordings of only 3 minutes, typical psychoacoustic elements and properties of the respective urban spaces are already captured.

In all, most of the measurements of the same location have relatively low distances over different years. Obviously, measurement location 06, called Katschhof, is very different from the other locations (dark grey). This location is a large urban square between the Rathaus (city hall) and cathedral without any commercial activities. This large urban square is surrounded by huge buildings and is acoustically shadowed. It describes an independent cluster.

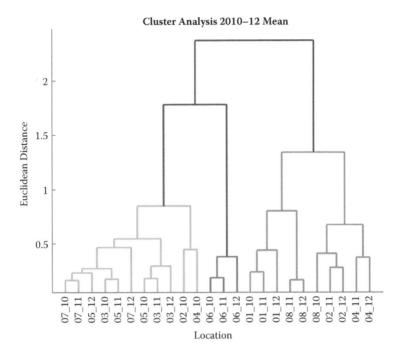

Figure 6.15 Cluster analysis of the measurements using the average linkage method. (Input data are z-transformed values of several per year and per location psychoacoustic parameters. The values of the two measurements per year were averaged.) The naming of the clustered objects is as follows: measurement location and, after an underscore, the year of measurement.

Another cluster (grey, right) consists of the locations 01 and 08, which are places with heavy road traffic very close to the road. The measurement places 02 and 04 are also close to the locations dominated by road traffic noise; however, these places are located near roads with a lower traffic volume. This is also the reason why these places measured in 2010 were assigned to the left cluster, since the measurements were performed in the evening and the traffic volume was very low then. The remaining places (03, 05, and 07) are combined to another major cluster (light grey, left). These places are urban squares with a lot of commercial activities (Marketplace and Elisenbrunnen Park) or a pedestrian zone with restaurants (Pontviertel) without significant road traffic noise. Obviously, these places are relatively similar in a psychoacoustic sense. To study the differences of the diverse locations, the analyzed measurements were subject to a principal component analysis.

Figure 6.16 shows the result of a principal component analysis, where the measurement locations are displayed with respect to the determined components 1 and 2, which together explain over 90% of variance. The loadings regarding component 1 correlate well with *intensity-related parameters*,

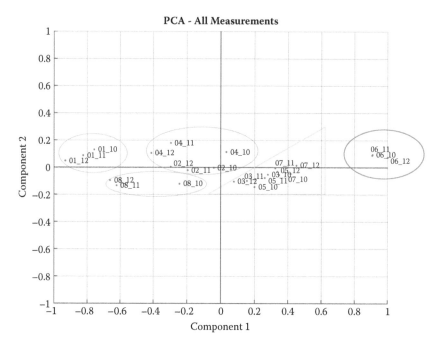

Figure 6.16 Principal component analysis: the loadings of the different measurements over the years with respect to components 1 and 2 are shown. The naming of the objects is as follows: measurement location and, after an underscore, the year of measurement.

such as loudness, sound pressure level, or articulation index. Component 2 is more related to *spectral aspects* correlating, for example, with sharpness. Similar to the cluster analysis result, the corresponding measurements over the years are very close together. Obviously, location 06 (Katschhof) is far away in a psychoacoustic perspective from the other locations. Locations 01 and 08 are assigned to the other extreme of component 1. However, 01 and 08 are different with respect to component 2, which is more related to the spectral content. Due to a fountain close to the measurement point 08, besides the road traffic noise, high-frequency noise is also present. This leads to higher sharpness values at position 08 than at position 01, which are both dominated by road traffic noise (see exemplarily Figure 6.17). Locations 02 and 04, where lower road traffic volume is present and pedestrian noise is also audible, are summarized in another cluster. Finally, locations 03, 05, and 07 (without road traffic noise and with a lot of commercial activities) are close together and show the closest proximity to location 06.

Although this statistical approach of identifying similarities between acoustical properties of acoustical environments is statistical and neglects other moderators, such as visual elements and context, it already gives first indications with respect to a potential classification of soundscapes.

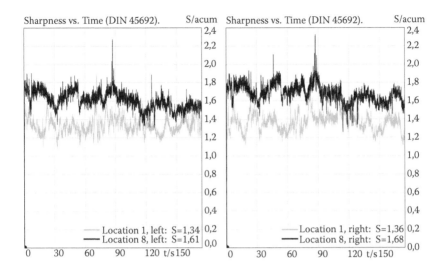

Figure 6.17 Sharpness analyses of measurements performed at locations 01 (grey) and 08 (black) according to DIN 45692. Left: Left channel. Right: Right channel.

In particular, it demonstrates that even short-term recordings of only a few minutes include typical acoustic elements, which can be grasped and described by psychoacoustic parameters to a certain extent.

Figure 6.18 illustrates that the link between the judgements of loudness and unpleasantness depends on the respective soundscape. The participants gave ratings in the soundwalk after listening for 3 minutes to the noise of the respective location. This time interval was also recorded and subject to psychoacoustic analysis. In the Templergraben case (04), where road traffic noise dominates the acoustic environment, the judgements of "How loud is it here?" and "How unpleasant is it here?" are very similar without any offset. In contrast, the judgements of loudness and unpleasantness show systematic differences in the Marketplace case (05). The participants consistently rate the loudness higher than the unpleasantness. In general, it turned out that the judgements of the different locations were comparable over the years (Fiebig, 2013). This observation is surprising. Although the sounds as well as the soundwalk participants varied over the years, comparable assessments of loudness and unpleasantness were provoked during the soundwalks.

This observation demonstrates the need for a soundscape approach: the existence of noise (e.g., expressed by level indicators) does not automatically indicate the level of unpleasantness. The sensations of loudness and unpleasantness cover different perceptual dimensions; the determination of the sound pressure level or even loudness does not describe the pleasantness of a soundscape. The simple assumption usually applied by noise regulations—loudness equals annoyance—is very misleading.

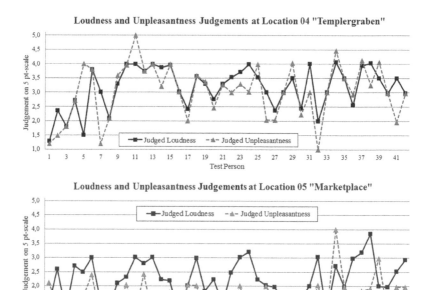

Figure 6.18 Loudness and unpleasantness judgements of locations 04 and 05 displayed over participants. Top: In situ judgements given at Templergraben. Bottom: In situ judgements given at Marketplace.

6.4 CHARACTERIZATION OF SOUNDSCAPES

6.4.1 Options and Limitations of Psychoacoustic Analysis

The consideration of psychoacoustic aspects, taking into account the human (binaural) signal processing, allows for identifying pleasant and unpleasant sound features from the psychoacoustic point of view. "From acoustics and psychoacoustics we will learn about the physical properties of sound and the way sound is interpreted by the human brain" (Schafer, 1977, p. 4). The most simple but important first step is to change from averaging energy descriptions (based on SPL values) to a more detailed acoustic description taking into account acoustic properties of the sound events. To achieve an improved description of the acoustic environment of a soundscape, psycho-acoustic parameters have to be applied. The use of psychoacoustic param-eters and the detection of temporal and spectral patterns will advance soundscape evaluations and will considerably improve perceptually related assessments of the environmental sound quality and its expected impact on annoyance (Genuit and Fiebig, 2006). New parameters, like the Relative Approach, must be developed to characterize the noise situation further.

Moreover, the idea of creating psychoacoustic maps (Genuit et al., 2008) integrating further parameters in conventional noise maps will allow, as a starting point, the application of the soundscape approach with psycho-acoustics to the environmental and community noise context. For more information see Chapter 7.

Of course, the presented technical–psychoacoustic approach for analyzing acoustical environments does not completely cover the investigation of the sensational and mental representation of urban space with its typical sound. Without asking how residents feel about their surroundings, without experiencing the place with its visual elements, a psychoacoustic analysis is almost meaningless. Aspects like local expectations, suitability, or acceptability of sound in context cannot be sufficiently answered without knowledge of human responses to considered places. Psychoacoustics can analyze in detail the acoustic composition of a soundscape, but for interpreting this composition correctly, the feedback from listeners is indispensable. Nevertheless, it is inevitable to thoroughly study the acoustical properties and psychoacoustic characteristics of soundscapes as an important part of the soundscape phenomenon, because it can provide a starting point for the classification of soundscapes. Several acoustical properties were identified that are typical for the examined urban locations. Acoustic particularities and noise features within different soundscapes have to be identified. In all, macroscopic and microscopic analyses are needed, on the one hand, to capture the "global" sound impression created by the soundscape and, on the other hand, to recognize and interpret single noise events, which cause strong reactions and feelings, whether positive or negative.

6.4.2 Meaning of Source Recognition and Listening Focus

The overall noise with respect to a specific receiver point can be measured and analyzed in terms of several acoustical parameters. However, the annoyance or pleasantness level of a complex soundscape containing different sound sources cannot be simply determined on the basis of single acoustical parameters. It is expected that due to specific source attention processes, the influence of a certain sound source can be very high, although the acoustic contribution to the overall noise caused by this source is not very important in a physical sense. Thus, it is necessary to study typical attention processes and the focus of the listener on sound sources in complex environments to understand the perception and evaluation of soundscapes.

As already mentioned in the sections above, due to human binaural listening and complex postprocessing realizing source focusing (improved signal-to-noise ratio) combined with noise suppression, the spatial distribution of sound sources, as well as the direction and speed of any movement of these sources, can be relevant for the perception and evaluation of environmental noise.

Since (binaural) signal processing involved in human hearing has to be considered, the use of aurally accurate measurements is mandatory and monaural measurements are not adequate.

Of course, the evaluation by a listener of the surrounding soundscape is also dependent on the personal attitude of the listener, which further increases the intricacy of the soundscape approach. Source meaning and connotation are particularly important, in case a sound source is dominating the soundscape or is exceedingly attracting attention. For more information see Chapter 2.

6.4.3 Outlook

Psychoacoustic parameters greatly cover several sensory dimensions describing the general sound character of (environmental) noises beyond doubt. Based on detailed knowledge of the psychoacoustic properties of noise, conclusions can be drawn with respect to the general character of the sound. However, to reliably derive the pleasantness and acceptability of the sound, more parameters are needed, going beyond classical psychoacoustic parameters. Analyses and indicators are needed that provide information about the *harmony* perception of a sound, whether heard tonal components are causing perception of harmony and consonance or dissonance. Indicators have to be developed, which combine musical understanding with psychoacoustic knowledge.

Another important research issue concerns *automatic source recognition*, which is imperative for a comprehensive soundscape investigation. Due to complex attention processes in multisource scenarios, it is very important for subsequent analyses to have detailed knowledge about the sound source constellation. Moreover, the specific meaning of sound sources within a soundscape influences greatly the appraisal of the soundscape. Therefore, the relevant sound sources must be determined. This is not a question of relevance in terms of their sound pressure levels, because humans can focus on even low-level sound sources by using their binaural signal processing. Here, future research is needed to improve options in the field of (blind) automatic sound source recognition.

REFERENCES

ANSI S12.9. Quantities and procedures for description and measurement of environmental sound. 2003.

W. Aures. Berechnungsverfahren für den Wohlklang beliebiger Schallsignale. Ein Beitrag zur gehörbezogenen Schallanalyse. Dissertation, München, Germany, 1984.

J. Blauert. *Spatial Hearing: The Psychophysics of Human Sound Localization.* Cambridge: MIT Press, 1996.

G. v. Bismarck. Sharpness as an attribute of the timbre of steady state sounds. *Acustica* 30, 157–172, 1974.

W. R. Bray. Behavior of psychoacoustic measurements with time-varying signals. Presented at Noise-Con 2007, Reno, NV, 2007.

DIN 45631/A1. Calculation of loudness level and loudness from the sound spectrum—Zwicker method—Amendment 1: Calculation of the loudness of time-variant sound. Beuth Verlag, 2010. Berlin, Germany.

D. Dubois, C. Guastavino, M. Raimbault. A cognitive approach to urban soundscapes: Using verbal data to access everyday life auditory categories. *Acta Acustica united with Acustica* 92(6), 865–874, 2006.

EU Project IMAGINE (Improved Methods for the Assessment of the Generic Impact of Noise in the Environment). Determination of L_{den} and L_{night} using measurements. IMA32TR-040510-SP08, 8th draft, 2006.

H. Fastl, E. Zwicker. *Psychoacoustics. Facts and Models*, Heidelberg, New York, Berlin: Springer Verlag, 2007.

A. Fiebig. Psychoacoustic evaluation of urban noise. Presented at Internoise 2013, Innsbruck, Austria, 2013.

A. Fiebig, V. Acloque, S. Basturk, M. Di Gabriele, M. Horvat, M. Masullo, R. Pieren, K. S. Voigt, M. Yang, K. Genuit, B. Schulte-Fortkamp. Education in soundscape: A seminar with young scientists in the COST short term scientific mission soundscape: Measurement, analysis, evaluation. Presented at Proceedings of the 20th International Congress on Acoustics (ICA 2010), Sydney, Australia, 2010.

A. Fiebig, K. Genuit. Applicability of the soundscape approach in the legal context. Presented at DAGA 2011, Düsseldorf, Germany, 2011.

A. Fiebig, S. Guidati, A. Goehrke. The psychoacoustic evaluation of traffic noise. Presented at NAG/DAGA 2009, Rotterdam, Netherlands, 2009.

IEC 959. Provisional head and torso simulator for acoustic measurements on air conduction hearing aids. Technical report. International Electronical Commission, 1990.

ISO 1996. Acoustics—Description and measurement of environmental noise. Part 1—Basic quantities and procedures, Part 2—Acquisition of data pertinent to land use, Part 3—Application of noise limits.

K. Genuit. Standardization of binaural measurement technique. *Journal de Physique IV, Colloque Ce, Supplément au Journal de Physique III*, 2, 1992a.

K. Genuit. *Sound Quality, Sound Comfort, Sound Design—Why Use Artificial Head Measurement Technology?* Stuttgart: JRC, Daimler-Benz, 1992b.

K. Genuit. Objective evaluation of acoustic-quality based on a relative approach. Presented at Inter-Noise 1996, Liverpool, England, 1996.

K. Genuit. Beyond the A-weighted level. Presented at Internoise 2006, Honolulu, HI, 2006.

K. Genuit, J. Blauert, M. Bodden, G. Jansen, S. Schwarze, V. Mellert, H. Remmers. *Entwicklung einer Messtechnik zur physiologischen Bewertung von Lärmeinwirkungen unter Berücksichtigung der psychoakustischen Eigenschaften des menschlichen Gehörs.* Dortmund: Schriftenreihe der Bundesanstalt für Arbeitsschutz und Arbeitsmedizin, Forschung Fb 774, Wirtschaftsverlag NW, 1997.

K. Genuit, A. Fiebig. Psychoacoustics and its benefit for the soundscape approach. *Acta Acustica united with Acustica* 92(6), 2006.

K. Genuit, A. Fiebig. Environmental noise: Is there any significant influence on animals? *Journal of the Acoustical Society of America* 122, 3082, 2007a.

K. Genuit, A. Fiebig. The acoustic description of patterns in soundscapes. Presented at Internoise 2007, Istanbul, Turkey, 2007b.

K. Genuit, W. Bray. Dynamic acoustic measurement techniques considering human perception. Presented at ASME International Mechanical Engineering Congress and Exhibition, Chicago, Illinois, USA, 2006.

K. Genuit, A. Fiebig. The measurement of soundscapes—Is it standardizable? Presented at Internoise 2014, Melbourne, Australia, 2014.

K. Genuit, B. Schulte-Fortkamp, A. Fiebig. Psychoacoustic mapping within the soundscape approach. Presented at Internoise 2008, Shanghai, China, 2008.

R. P. Hellman, E. Zwicker. Why can a decrease in dB(A) produce an increase in loudness? *Journal of the Acoustical Society of America* 2(5), 1987.

G. Jansen, P. Rey. Der Einfluss der Bandbreite eines Geräusches auf die Stärke vegetativer Reaktionen, Internationale Zeitschrift für Angewandte Physiologie. *Einschließlich Arbeitsphysiologie* 19, 209–127, 1962.

M. A. Nobile, G. R. Bienvenue. Procedure for determining the prominence ratio of discrete tones in noise emissions. Presented at Noise-Con, Tarrytown, NY, 1991.

R. M. Schafer. *The Soundscape: Our Sonic Environment and the Tuning of the World*. Rochester, VT: Destiny Books, 1977.

B. Schulte-Fortkamp, A. Fiebig. Soundscape analysis in a residential area: An evaluation combining noise and people's mind. *Acta Acustica united with Acustica* 92(6), 2006.

R. Sottek. Modelle zur Signalverarbeltung im menschlichen Gheör. Dissertation, Aachen, Germany, 1993.

I. M. Stemplinger. Beurteilung, Messung und Prognose der Globalen Lautheit von Geräuschimmissionen. Dissertation, München, Germany, 1999.

M. Vorländer. Auralization: *Fundamentals of Acoustics, Modelling, Simulation, Algorithms and Acoustic Virtual Reality*. Berlin: Springer Verlag, 2008.

Chapter 7

Mapping of Soundscape

Jian Kang,[1] Brigitte Schulte-Fortkamp,[2]
André Fiebig,[3] and Dick Botteldooren[4]

[1]University of Sheffield, Sheffield, United Kingdom

[2]Technical University of Berlin, Berlin, Germany

[3]HEAD acoustics, Herzogenrath, Germany

[4]Ghent University, Ghent, Belgium

CONTENTS

Mapping is a useful tool to aid the design and planning process. For strategic planning at a larger spatial scale, the European Environmental Noise Directive (END) requires Lden and Lnight maps to be drawn. However, these only consider noise sources, especially traffic noise sources, and they are often not well related to people's perception of the acoustic environment. This chapter presents some mapping techniques based on recent research, including sound mapping, which shows sound-level distribution considering more source types, both positive and negative, rather than just traffic noise; soundscape mapping based on human perception of sound sources; soundscape mapping developed using artificial neural networks (ANNs), which show people's perception; psychoacoustic mapping and mind mapping; and mapping of noticed sounds. There are also a number of other soundscape mapping techniques, such as soundscape topography (Boubezari et al., 2011; Boubezari and Bento Coelho, 2012). The chapter also describes how a map of sounds that are likely to be noticed by the users of the space can be constructed from the knowledge of the sonic environment as a tool both for understanding the soundscape composition and for design.

7.1 MAPPING SOUND FIELD

A sound map, typically in a form of interpolated isocontours, is a way of presenting geographical distribution of sound exposure, in terms of either measured or calculated levels. While large-scale, multiple-receiver in situ measurement is not always feasible, and most acoustic problems cannot be resolved purely analytically, sound mapping is often based on computer simulation of the sound field, although another relevant approach is physical scale modelling, for relatively small-scale urban areas. Compared with in situ measurements in urban areas, an advantage of computer modelling is that the geometry, source and receiver condition, and background noise can relatively easily be modified.

7.1.1 Microscale Sound Field Mapping

At microscale, such as a street or a square, sound field mapping is often based on three-dimensional (3D) computer simulation techniques. The *image source method* treats a flat surface as a mirror and creates an image source. In other words, the boundaries are regarded as geometrically (specularly) reflective. The reflected sound is then modelled with a sound path, directly from the image source to a receiver. Multiple reflections are achieved by considering further images of the image source. For each reflection, the strength of the image source is reduced due to surface absorption (Kang, 2000, 2002a). A disadvantage of the image source method is that the calculation speed is reduced exponentially with increasing orders of reflection as the number of images increases. In addition, validity and visibility tests are required for image sources.

The *radiosity method* provides an efficient way to consider diffusely reflecting boundaries. The method functions by dividing boundaries, in a space such as an urban street or square, into a number of patches (i.e., elements) and replaces the patches and receivers with nodes in a network. The sound propagation within the space can then be simulated by energy exchange between those nodes. Various computer programs have been developed based on the radiosity method that are applicable to urban spaces such as street canyons and urban squares (Kang, 2001, 2002b, 2005). A model combining ray tracing and radiosity has also been developed and well validated against measurements (Meng et al., 2008).

Figure 7.1 shows the sound distribution of a single point source (e.g., a car, a speaker, or a fountain) at five different positions in an idealized cross street, based on calculation using the radiosity method, where the street boundaries are assumed to be diffusely reflective and have an absorption coefficient of 0.1. Figure 7.2 shows a sound map in a typical urban square, the Peace Gardens in Sheffield, UK, where the fountain, which was positively evaluated by users, and traffic noise from the road are both considered (Yang and Kang, 2005a, 2005b).

Not only sound level, but also reverberation time is important in terms of soundscape perception. Microscale acoustic simulation techniques as described above can also provide reverberation. Figure 7.3 shows the simulated reverberation time RT30 in a typical urban square, Piazza della Signoria in Florence, Italy, where a point source is assumed at the middle of the square.

With the development of more powerful computers, a number of other models, based on numerically solving the wave equations, have also been developed and applied in urban situations (Kang, 2007). The most popular of these are the finite difference time domain (FDTD) method and its cousin the pseudospectral time domain (PSTD) method (Van Renterghem et al., 2013). The equivalent sources method (ESM) and parabolic equation (PE) method are mainly used to include propagation above the roofs (Van Renterghem et al., 2006). Finally, finite element (FEM) and boundary element (BEM) methods have also been explored.

Moreover, to aid urban soundscape design, as well as public participation, it is useful to present the 3D visual environment with an acoustic animation/auralization tool, where consideration is given to various urban sound sources, dynamic characteristics of the sources, and movements of sources and receivers. The calculation speed should be reasonably fast, so that a designer can adjust the design, listen to the difference, and create an instant evaluation (Smyrnova and Kang, 2010).

7.1.2 Macroscale Sound Field Mapping

Although microscale simulation techniques, as described above, can generate relatively accurate predictions of urban sound propagation,

Figure 7.1 Microscale sound map—sound distribution of a point source in an idealized cross street. (From Kang, J., *Urban Sound Environment*, London: Taylor & Francis, 2007.)

it is generally inappropriate to apply these algorithms at a macroscale, such as a relatively large urban area or the whole city. At macroscale, sound field mapping normally involves statistical methods and simplified algorithms, for example, applying a 2.5-dimensional approach, which means that all structures are vertically erected on a 2D map. A number of software packages have been developed for large-area noise mapping (Kang, 2007).

A common feature of all noise mapping software is the combination of noise propagation calculations with a mapping and scheme editing facility, consisting of georeferenced input data, often associated with geographical information systems (GISs). Building a model is the next important

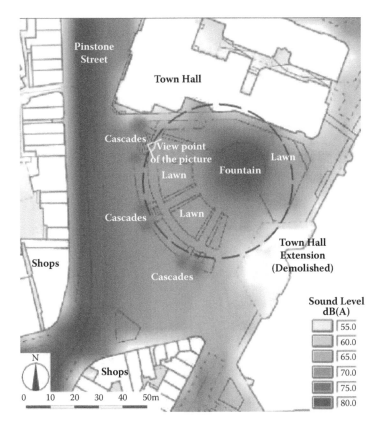

Figure 7.2 Sound map in the Peace Gardens in Sheffield, UK. (From Yang, W., Kang, J., *J. Urban Design,* 10, 69–88, 2005.)

step. Depending on the scale of the terrain in terms of vertical differences, ground elevation data are normally required at 5 m intervals. Positional and height information is also required for buildings and any major structures. This information can be obtained either from maps, from aerial photographs, or by survey.

Positions and characteristics of various types of sound source are also needed. Much effort has been made to develop source models for road traffic. Noise source points on vehicles are simplified by adopting just two point sources: the first, a low source height at 0.01 m above the road surface, which typically represents tyre/road noise, and the second, a higher source point, which largely reflects propulsion noise (e.g., engine), with height dependent on the vehicle category. Other sound sources can also be taken into account, for example, fountains and birdsongs, if the source characteristics are given, including sound power level and directionality.

The calculation of sound propagation is based on a series of simplified algorithms, specified in various standards, internationally (ISO, 1993)

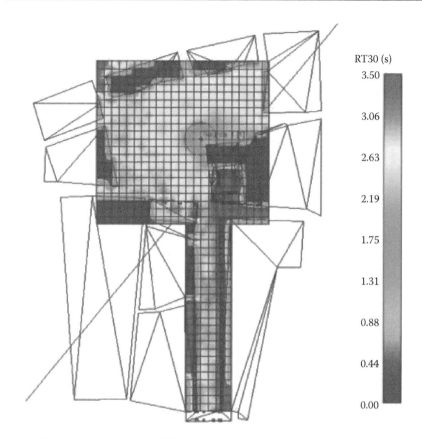

Figure 7.3 Reverberation time RT30 of the simplified model of Piazza della Signoria in Florence, Italy. (From Yang, W., An aesthetic approach to the soundscape of urban public open spaces, PhD thesis, School of Architecture, University of Sheffield, Sheffield, UK, 2005.)

and nationally, considering various sources, including aircraft, road, railway, and industry. In ISO 9613, the equivalent continuous downwind octave-band sound pressure level (SPL) at a receiver position is determined by considering the sound power level of the source, a directivity correction, and the attenuation between source and receiver, which includes the geometrical divergence, atmospheric absorption, ground effects, barrier effects, and miscellaneous effects. Appropriate calculation parameters should be chosen, such as reflection order and the radius within which sources could affect the SPL, evaluation parameters and reference time periods, and grid factors for dividing line or area sources.

After the calculation process, noise maps can be produced, either horizontally above the ground or vertically in front of building façades, as well as other outputs, such as exposure levels of a population for risk estimation purposes.

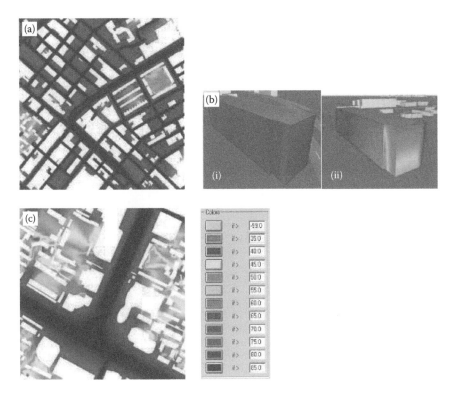

Figure 7.4 Macroscale sound field mapping for traffic noise. (a) A typical urban area in Greater Manchester, UK, plan view. (b) A given building at two different locations in Greater Manchester, UK, façade view. (From Barclay et al., *Build. Environ.*, 52, 68–76, 2012.) (c) A typical urban area in Wuhan, China, plan view. (From Wang, B., Kang, J., *Appl. Acoust.*, 72, 556–568, 2011.)

Figure 7.4 shows two examples of sound maps considering traffic noise in Manchester, UK, and Wuhan, China, and different urban morphology (Wang and Kang, 2011). Although under the EU END requirements (2002), extensive work has been carried out on noise mapping, for the identification and protection of quiet areas, it is also important to take into account more sound types, including positive sounds. Figure 7.5 is a map showing the distribution of bird sound in a residential area (Hao et al., 2013), where it is assumed that birds are located on the edges of the green areas. By overlapping the sound maps of different sound sources, masking effects can be examined.

7.2 SOUNDSCAPE MAPPING BASED ON HUMAN PERCEPTION OF SOUND SOURCES

In soundscape research, soundscapes were usually considered the full range of perceptible sounds in a given landscape at a given time (Liu et al., 2013,

2kHz   4kHz   8kHz

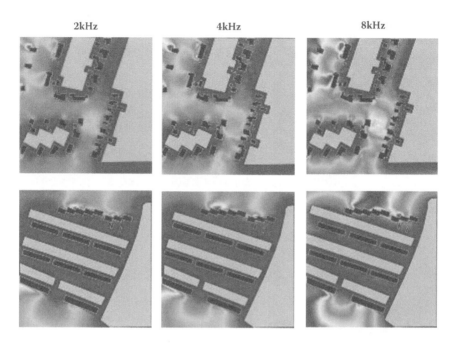

Figure 7.5 Distribution of bird sound in a residential area. (From Hao, Y., et al., *Landsc. Urban Plan.*, accepted, 2013.)

2014b). It would be useful to visualize the spatial distribution of certain sound sources and the soundscape spatiotemporal dynamic in a relatively large area for potential soundscape information users. Thus, thematic soundscape maps could be developed as additional layers of landscape information. The hypothesis of developing these maps is that based on the soundscape information on certain locations in an area, the soundscape of the whole area could be predicted with the spatial analysis method in GIS software.

The mapping process will be explained by introducing a case study in northeast Germany, in the Warnemünde district of Rostock on the Baltic Sea. The study area extends almost 2400 m east–west and 2000 m north–south, as shown in Figure 7.6 (Liu et al., 2013, 2014a, 2014c).

7.2.1 Preparation for the On-Site Investigation

A group of 12 observers without hearing deficiencies (7 male and 5 females; aged between 22 and 31; mean, 26; standard deviation, 2.8) were recruited to collect the soundscape data. During a pilot study, major sounds that frequently appear in this area were recognized and coded, as shown in Table 7.1, as a reference for the observers. Twenty-three approachable sites were evenly sampled across the study area (Figure 7.6), based on fishnet cells of 350*350 m created in ArcGIS 9.1 (Matsinos et al., 2008), to ensure a representative soundscape mapping of the whole area.

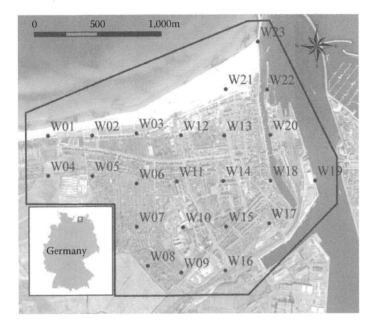

Figure 7.6 Location of the study area and distribution of the 23 sampled sites (W01 to W23).

7.2.2 Quality Control Process

All the observers participated in a training process 1 month before the on-site survey, including getting familiar with the list of sounds through watching videos recorded on site and making field practice to control for observation bias, in order to guarantee a consistent and comparable evaluation of soundscapes. In other words, the training was not to make the observers expert listeners, but rather to make sure they understood what was required to evaluate as common listeners, so that comparable results could be obtained. An interrater reliability test was also conducted to control the evaluation process (Cronbach's alpha value = 0.91). In order to increase recording efficiency, the observers were divided into six groups of two observers, respectively (Liu et al., 2014a).

7.2.3 Soundscape Data Collection

The investigation was carried out on August 3–4, 2011. Data were collected in eight 2 h successively sampled periods at each site:

First period: 0600–0800
Second period: 0800–1000
Third period: 1000–1200

Table 7.1 Classification of Major Sound Categories in the Study Area

Sound Category	Major Sound	Code
Human sound	Child voice	CS
(Hum)	Adult voice	AS
	Footstep	FS
Mechanical sound	Airplane flying	AF
(Mech)	Bicycle riding	BC
	Bell ringing	BR
	Construction	CT
	Emergency	ES
	Grass mowing	GM
	Music	MS
	Ship moving	SM
	Other mechanical sounds	OA
Traffic sound	Train moving	TM
(Traf)	Traffic sound (foreground)	TSF
	Traffic sound (background)	TSB
	Motorcycle rumbling	MR
Geophysical sound	Grass rustling	GR
(Geo)	Raining	RS
	Sea wave	SW
	Tree rustling	TR
	Wind blowing	WF
	Water sound	WS
Biological sound	Birdsong	BS
(Bio)	Chicken clucking	CC
	Dog barking	DB
	Frog	FR
	Insects	IS

Fourth period: 1200–1400
Fifth period: 1400–1600
Sixth period: 1600–1800
Seventh period: 1800–2000
Eighth period: 2000–2200

Within each sampled period, a 10 min period of soundscape was randomly recorded. Each 10 min period was further divided into 20 sequential time steps of 30 s. Within each time step, the recognized sounds were recorded with code names and their perceived loudness was scored accordingly, by using a 5-point linear scale (1 = very quiet, 2 = quiet, 3 = normal, 4 = loud,

5 = very loud). For each sound, the score was given according to the highest one during the time step. Any sound that did not appear in a given time step was categorized as 0. The accumulated perceived loudness of a certain sound source at a given site and period was calculated by adding the scores obtained from the 20 sequential time steps (30 s) of the 10 min period. Similarly, perceived loudness of each sound category was the sum of scores of all the corresponding sound sources.

7.2.4 Thematic Soundscape Mapping

Thematic soundscape maps were generated using a regularized spline with a tension interpolation method implemented in the GRASS software (GRASS Development Team, 2008), since it can generate a smoother surface than ArcGIS. Grid maps of each sound source/category could be laid across the whole study area based on their respective accumulated loudness at each sampled site per period or the whole day.

Consequently, soundscape information with spatiotemporal characteristics could be visually presented above the underlying landscape. It needs to be noted that with a different spatial and temporal sampling method, the mapping results could be different for the same area.

7.2.4.1 Mapping of Single Sound Source

As an example, the mapping results of spatial distribution of perceived loudness of birdsong in each sampled period are shown in Figure 7.7 (Liu et al., 2014a). All the maps are presented using the same scale (0–100), in order to make them easily comparable by the colour. It can be seen that the distributions of birdsong across the study area showed an ever-changing characteristic along with different sampled periods. However, a clear spatial pattern was shown in each period, that is, there was always relatively more birdsong perceived at certain sampled sites than others. Higher perceived loudness of birdsong was normally concentrated in residential areas (W14, W17, W11, W12), garden areas (W05, W06, W07, W08), and urban park (W13).

7.2.4.2 Mapping of Certain Sound Category

Figure 7.8 shows the mapping results of the five sound categories, that is, human, mechanical, traffic, geophysical, and biological sounds, respectively (Liu et al., 2014c). It can be seen that more human sounds appeared across the beach area, as a lot of local people and tourists concentrated in this area. More mechanical sounds were concentrated along the southeastern and eastern boundary of the study area, where the railway passes through and construction work was carried out at site W16. It is obviously that traffic sounds concentrated almost along the direction of W09–W15, where the widest traffic road in the study area passes through. Geophysical

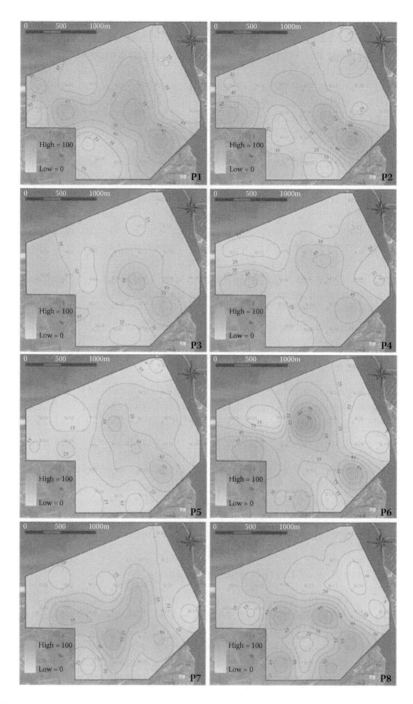

Figure 7.7 Perceived loudness of birdsong across the study area during the first to eighth sampled periods, respectively (map P1–P8).

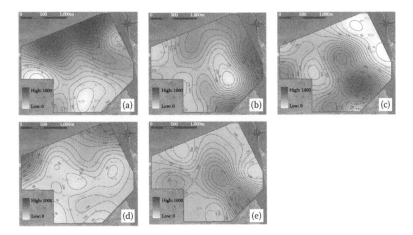

Figure 7.8 Daily accumulated perceived loudness of human (a), mechanical (b), traffic (c), geophysical (d), and biological (e) sounds across the study area.

sounds were mainly perceived along the beach area, especially on the two ends, because the sea waves and wind were stronger. More biological sounds appeared in the central and southeastern parts of the study area, mainly because dense constructions of the residential area form a quiet environment, thus preventing fragile biological sounds from masking by other sounds, and dense vegetation of the urban park is ecologically good habitat for vocalizing organisms such as birds and insects.

7.2.4.3 Mapping of Soundscape Composition

If the sound sources are reclassified into three major categories, that is, anthrophony, biophony, and geophony, and their grid maps are represented by red, green, and blue, respectively, these maps could be combined in ArcGIS 9.1 to show what the soundscape composition looks like. This is based on simulating the compounding principle of the three original colours to other intermediate colours that stand for areas that perceived combined sounds. Figure 7.9 shows the soundscape composition maps of the first four periods. The strongly varying colours seen in these maps indicate complex spatiotemporal patterns of the soundscape composition, and also a clear dominating position of anthrophony.

7.3 SOUNDSCAPE MAPPING BASED ON ANN

For urban planners and architects, it would be useful to develop a tool to predict the subjective evaluation of soundscape quality by potential users, using known design conditions such as physical features of a space, acoustic

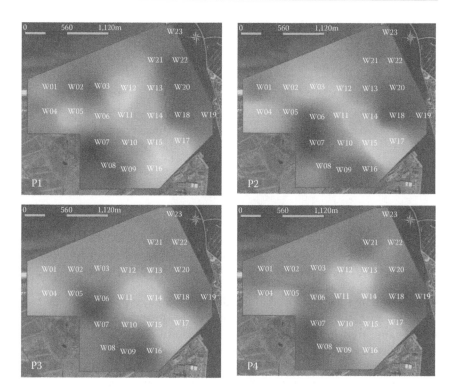

Figure 7.9 Soundscape composition during the first four sampled periods, where anthrophony, biophony, and geophony are originally described in red, green, and blue, respectively. Map P1 to P4 describes the soundscape composition during the first through fourth sampled periods, respectively.

variables, and people's characteristics. In order to develop the computer models, the artificial neural network (ANN) technique has been introduced and employed. The hypothesis of the study is that a well-trained ANN model derived from existing sites can be used to predict subjective evaluation of soundscape in new urban open spaces with similar physical and social environments (Yu and Kang, 2009).

7.3.1 Affecting Factors for the Subjective Evaluation

The results of a series of field surveys have been used to gather the data for ANN models. The surveys were first undertaken in two urban open spaces in each of the following seven cities in Europe: Athens, Thessaloniki, Milan, Fribourg, Cambridge, Sheffield, and Kassel. Parallel surveys were carried out in five Chinese urban open spaces, two in Beijing and three in Shanghai. In total, more than 9000 people in Europe and more than 800 people in China were interviewed. The investigated factors in the field surveys are shown in Table 7.2. For the subjective evaluation of acoustic comfort, five scales were used, from –2 (very uncomfortable) to 2 (very comfortable).

In the following analysis, both sound-level evaluation and acoustic comfort evaluation are regarded as output, and the relating factors to each of them are examined. It is noted that for the acoustic comfort evaluation, the sound-level evaluation is also used as an input variable. This is because the former is regarded as the final outcome of soundscape quality, whereas the latter is one of its components.

As ANN has a robust learning capability to model nonlinear relationships, many input variables can be used in the network as long as there are sufficient training samples. However, since the sample sizes vary considerably among different case study sites, it is important to limit the input variables so that the network size can be kept reasonable for a good prediction. On the other hand, if the input variables are selected too strictly, namely, only those factors that are highly related to the output are used, the advantage of ANN modelling compared to a simple multiple regression would not be significant. The statistical analyses of the field survey results have been made using SPSS, considering the Pearson/Spearman correlations (two-tailed) and Pearson chi-square for factors with 3+ scales, namely, taking both linear and nonlinear correlations into account, as well as mean differences (t-test, two-tailed) for factors with two scales. Correspondingly, a series of affecting factors for the subjective evaluation of acoustic comfort and sound level have been derived. Overall, the affecting factors are rather different in various case study sites.

7.3.2 ANN Modelling

The learning process of ANN is to develop weights between its processing elements, including various input nodes, hidden nodes, and output nodes. The weights stress the response strength. In the whole training process, the weights are constantly adjusted to reduce the differences between desired and actual responses.

Qnet, a back-propagated multilayer forward neural modelling system, was used. The accuracy of prediction is determined by the training and test correlation coefficients, as well as the root-mean-square error (RMSE) considering both training and test. The correlation coefficients are calculated between the network outputs and the targets from actual data. The RMSE and also the root-mean-square deviation (RMSD) are to measure the differences between outputs and targets.

For both sound-level evaluation and acoustic comfort evaluation, the input variables include the subjective evaluations of physical conditions such as temperature, wind, humidity, and brightness, given the multiple relationships between various factors. While this is possible based on the

Table 7.2 Factors Considered in the Surveys

Attributes	Elements	Ref. No.	Attribute Factors	Measures of the Attributes
Explicit	Physical	Phy1	Season	1: Winter
				2: Autumn
				3: Spring
				4: Summer
		Phy2	Time of day	1: Night: >2100
				2: Evening: 1800–2059
				3: Morning: 900–1159
				4: Afternoon: 1500–1759
				5: Midday: 1200–1459
		Phy3	Air temperature	Measurement of air temperature: °C
		Phy4	Wind speed	Measurement of wind speed: $m.s^{-1}$
		Phy5	Relative humidity	Measurement of relative humidity: %
		Phy6	Horizontal luminance	Measurement of horizontal luminance: Klux (EU), lux (China)
		Phy7	Sun shade	0: Interviewee not standing in the sun
				1: Interviewee standing in the sun
		Phy8	Sound pressure level	Measurement of sound pressure level: dB(A)
	Behavioural	B1	Whether wearing earphones	0: Not wearing earphones
				1: Wearing earphones
		B2	Whether reading or writing	0: Neither reading nor writing
				1: Either reading or writing
		B3	Whether watching somewhere	0: Not watching anywhere
				1: Watching somewhere
		B4	Movement status	1: Sitting
				2: Standing
				3: Playing with kids
				4: Sporting
		B5	Frequency of coming to the site	Scale 1–5: 1 = first time, 5 = every day
		B6	Reason for coming to the site	1: Equipment/services of the site
				2: Children playing and social meetings
				3: Business/meeting/break
				4: Attending social events

Table 7.2 (Continued) Factors Considered in the Surveys

Attributes	Elements	Ref. No.	Attribute Factors	Measures of the Attributes
				5: Passing by
		B7	Grouping: whether accompanied	0: Without company
				1: With 1 person
				2: With more than 1 person
Implicit	Social/ demographic	S1	Age	1: <12
				2: 12–17
				3: 18–24
				4: 25–34
				5: 35–44
				6: 45–54
				7: 55–64
				8: >65
		S2	Gender	1: Male
				2: Female
		S3	Occupation	1: Students
				2: Working people
				3: Others (e.g., unemployed and pensioners)
		S4	Education	1: Primary
				2: Secondary
				3: High level
		S5	Residential status	0: Nonlocal
				1: Local
		S6	Sound-level experience at home	Scale −2 to 2, with −2 as very quiet and 2 as very noisy
	Psychological	Psy1	Site preference	0: Do not like the site for certain reasons
				1: Like the site
		Psy2	View assessment	Scale −1 to 1, with −1 as negative and 1 as positive
		Psy3	Heat evaluation	Scale −2 to 2, with −2 as very cold and 2 as very hot
		Psy4	Wind evaluation	Scale −2 to 2, with −2 as stale and 2 as too much wind
		Psy5	Humidity evaluation	Scale −2 to 2, with −2 as very damp and 2 as very dry
		Psy6	Brightness evaluation	Scale −2 to 2, with −2 as very dark and 2 as very bright
		Psy7	Overall physical evaluation	0: Not comfortable
				1: Comfortable
		Psy8	Sound-level evaluation	Scale −2 to 2, with −2 as very quiet and 2 as very noisy

field surveys in the model development in this study, if the ANN models are to be used at the design stage, those input variables will not be available. However, there have been established relationships between these physical conditions and their subjective evaluations, which can be used in the soundscape ANN models at the design stage.

To model the sound-level evaluation, a general model was first explored, using the data from all 19 case study sites, representing a variety of urban open spaces. According to the significance levels analyzed by combining all the case study sites into one data set, 16 factors were chosen as input variables, and a number of models using different hidden layers and nodes were constructed. However, none of them converged. This suggested that a general model including all kinds of urban open space was not feasible. Models were therefore developed based on individual case study sites. Four sites from Europe were randomly chosen, and encouraged by the good results of individual models, efforts were then made to develop models for urban open spaces with similar characteristics. The 19 case study sites were first classified as four types according to their locations/functions: 7 case study sites were located in city centres, 5 in residential areas, 5 at tourist spots, and 2 near railway stations. For each type some case study sites were grouped according to their city/country/continent. In total, eight models were developed. The best prediction result was achieved in the model with two Cambridge sites, which were both tourist spots. The test coefficient is 0.6. For the three city centre models, the test coefficients are 0.48, 0.52, and 0.45, respectively, which are also satisfactory. Overall, the results on the subjective evaluation of sound level suggest that a general model for all the case study sites is not feasible due to the complex physical and social environments in urban open spaces. Models based on individual case study sites perform well, but their application range is limited. Specific models for certain types of location/function could be reliable and also practical.

The acoustic comfort evaluation was made for seven case study sites. Similar to the model development for the sound-level evaluation, a general model was first explored. It was shown that the test coefficients are generally satisfactory, but are rather low for certain sites.

Models based on individual case study sites were then developed to further examine the prediction performance. The prediction results are rather good for both models, especially for the Peace Gardens model, where the test coefficient reaches 0.79 and the RMSE is only 0.103. Acoustic comfort evaluation models were then developed based on case study sites with similar locations/functions. Two models were built for the city centre locations, one for the two case study sites in Sheffield and the other for two case study sites in China. Both models have rather good prediction performance. For the Sheffield model, compared to the individual model of the Peace Gardens, the test coefficient becomes slightly lower, by 0.11, whereas for the Chinese model, compared to the individual model of the Xi Dan Square, the test coefficient is the same.

Compared to the sound-level evaluation models, the prediction performance of acoustic comfort models is considerably better. This might mainly be caused by the role of input variables.

7.3.3 Mapping

The above ANN models can predict the sound-level and acoustic comfort evaluation of individual receivers/zones in an urban open space, given that the physical factors, such as SPL and user profiles, could vary at different positions. Sound-level and acoustic comfort evaluation maps can be produced accordingly, which would be a very useful tool for urban planners and designers.

Figure 7.10 shows the prediction maps for the sound-level evaluation and acoustic comfort evaluation in the Peace Gardens in terms of two age groups, 13–18 and >65. The SPL of each cell shown in the figure, which was used in the ANN models as an input variable, was calculated using noise mapping software Cadna, with source conditions similar to those in the field survey. In other words, to a certain extent, the mapping results should be regarded as predictions for new situations, rather than representations of the current situation. Other conditions were assumed to be the same between the two age groups and were based on the overall situation in the field survey, so that the result in each cell could be regarded as the average evaluation of each age group. From Figure 7.10, it can be seen that the 13–18 age group will generally feel quieter than the >65 age group, whereas in terms of acoustic comfort, the evaluation of the two age groups is very similar.

7.4 PSYCHOACOUSTIC MAPPING AND MIND MAPPING

As mentioned above, since the soundscape concept is focused on the perceiving human being and a soundscape is a perceptual construct related to, but not identical with, a physical phenomenon, the so-called noise mapping of an acoustic environment alone is insufficient. Thus, soundscape mapping needs to not only incorporate information about people's perception, but also include people via participation in the process of related mapping. A necessary step into the direction of mind mapping is to comprehensively cover basic auditory sensations through psychoacoustics. Such auditory sensations represent the perception of indivisible properties of a sound and can be understood as bottom-up perception. A sound can lead to a multitude of auditory sensations, like sensation of loudness, sharpness, roughness, tonality, and so forth. Consequently, the psychoacoustic analysis of acoustic environments is necessary but still not sufficient. Cognitive levels of sound perception have to be considered.

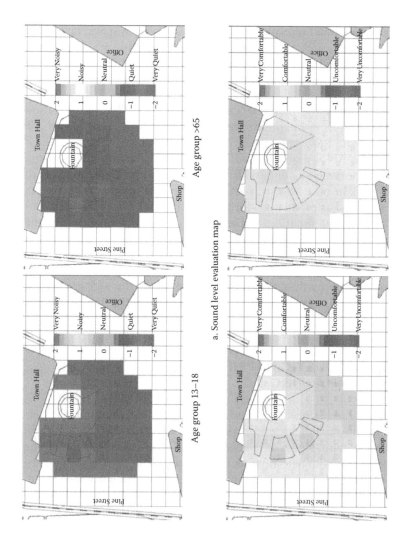

Figure 7.10 Prediction maps of sound-level evaluation and acoustic comfort evaluation in the Sheffield Peace Gardens (site 14). The calculated SPL is shown in each cell. Age group 13–18. Age group > 65. (b) Acoustic comfort evaluation map. (From Yu, L., Kang, J., *J. Acoust. Soc. Am.*, 126, 1163–1174, 2009.)

7.4.1 Multidisciplinary Access to People's Minds

Sound perception implies cognitive processing and interpretation of sound. Meaning is attributed to sound events in soundscapes; sounds evoke associations, constrictor support human activities, and sounds correspond to or are inconsistent with the perceiver's expectation. These processes are beyond the physical stimulus and must be explored by means of tools and methods of human and social sciences. In general, it is essential to apply a sociological approach for studying soundscape perception, because the whole process of perceiving and interpreting sound is based on social and cultural contexts strongly influencing the frame of reference (Schulte-Fortkamp and Fiebig, 2006). Thus, interview methods with varying degrees of structure and also soundwalks must be carried out to collect significant data. In this context, qualitative and open interviewing currently undergoes a methodological change and is considered to provide valid data. The interviewee is no longer seen as a distant, quantified, categorized, and catalogued faceless respondent, but has become a living human being (Kang et al., 2013). The measurement—the open interview (e.g., narrative, in-depth, or conversational interview)—will make an everyday conversation available for the participants. An emphatic understanding will provide reliable, rich, and deep data and also give room for the expression of emotional feelings. It is a conversation with a purpose, in which the interviewer aims to obtain the perspectives, feelings, and perceptions from the interview participants. Thus, based on such interview outcomes, a detailed picture of the soundscape as perceived by the people concerned can be drawn. It is evident that the interviewer must meet high demands on interview performance, avoiding interviewer effects and biases.

7.4.2 Nauener Platz, Berlin: Soundscape Study Increasing the Quality of Life for the People Concerned

To illustrate the process leading to mind mapping, the soundscape project Nauener Platz, Berlin, which received the European Soundscape Award in 2012, is presented. In Berlin, the public space Nauener Platz was rebuilt through participation of the people concerned, the so-called new experts (Schulte-Fortkamp et al., 2008, 2010). For determining the acoustic environment with its specific sound sources and their relevance for life and well-being, a methodology, including acoustic measurements, soundwalks, and open interviews, was applied. As a starting point, a conventional noise map was generated, providing general information about the noise exposure, as shown in Figure 7.11. The noise of two major roads with a high daily traffic volume acoustically dominated Nauener Platz.

Therefore, this map will not provide any information about the relationship of the neighbourhood community together with their set of expectations

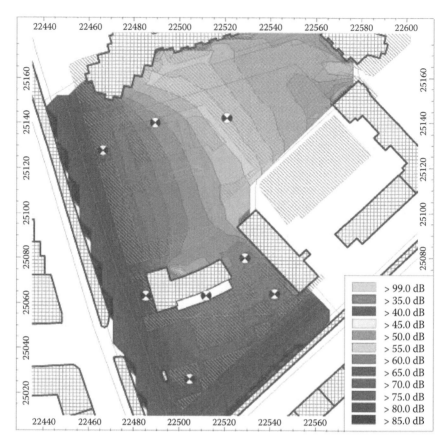

Figure 7.11 Noise map of public space Nauener Platz in Berlin. (From Schulte-Fortkamp, B., et al., Developing a public space in Berlin based on the knowledge of new experts, presented at Internoise 2008, Shanghai, China, 2008.)

concerning the acoustic environment. Moreover, it does not offer any details about responses and outcomes of residents toward their acoustic environment and existing frame of references leading to certain behaviour, attitudes, and coping strategies. Figure 7.12 shows the benefit and limitations of psychoacoustic maps in the context of soundscape investigations. Based on the loudness, sharpness, and roughness maps, the psychoacoustic properties of public space can be explained.

It is very important to consider the complex human cognitive stimulus integration over time with respect to the displayed psychoacoustic parameters. For example, it is known that prominent and loud noise events dominate the overall loudness sensation and must be emphasized with respect to the physical representation of loudness. Thus, the loudness value is shown, which is only exceeded in 5% of the measurement time (the so-called 5th percentile loudness). Such cognitive stimulus integration processes must

Figure 7.12 Psychoacoustic noise maps of public space Nauener Platz in Berlin. Schematic distribution of loudness (a), sharpness (b), and roughness (c) over the investigated area. (From Genuit, K., et al., Psychoacoustic mapping within the soundscape approach, presented at Internoise 2008, Shanghai, China, 2008.)

be considered regarding the determination of single values representing the sensation. The maps created in Figure 7.12 are only approximations to the actual conditions; the distribution of the different psychoacoustic parameters is estimated on the basis of the measured values.

In contrast to the characteristics of the conventional level or even loudness map, the psychoacoustic parameters sharpness and roughness do not exhibit a comparable behaviour. Sharpness considers the spectral shape of a noise, and roughness calculation is a nonlinear modulation analysis with a specific weighting with respect to frequency and modulation rate. It evaluates specific modulation characteristics in signals. The sharpness and roughness values are relatively constant all over the place. The reaction to the acoustic event will be related to pattern recognition.

In summary, if it is assumed that further acoustic properties besides the sound pressure level contribute to the assessment of the acoustic

Figure 7.13 Data collection (soundwalk) for mind mapping.

environment, then a prediction based on sound pressure level alone is inaccurate and is different than a prediction of perception using psychoacoustic parameters. For example, certain psychoacoustic properties are quite constant over the investigated area due to their (almost) independence from the absolute sound pressure level.

For the determination of the perception of the soundscape, it is necessary to focus on the people concerned. Based on soundwalks and interviews, it is possible to sufficiently learn about the process of perception and evaluation, as they take into account the context, ambiance, usual interaction between noise and listener, and multidimensionality of sound perception (see Figure 7.13) (Schulte-Fortkamp et al., 2008).

Using the soundwalk method, it turned out that all involved groups defined identical relevant listening places. Analyzing the comments and group discussion and taking into account the results from the narrative interviews, it was explored why people prefer some places over public space and why not. It also became clear how people experience the noise in the distance from the main roads. As an outcome of the interviews and listening test with more than 80 people living in the area, it became obvious that the most wanted sounds in this area are based on needs to escape road traffic noise through natural sounds. People requested "green acoustics," like singing birds or water sounds, which would allow mental escape from the omnipresence of road traffic noise. Consequently, it turned out that it is not the quietness of an area, but something that can be described as a need for harmony based on compatibility, assimilation, and acoustical home related to the respective expertise of the involved people. This provides knowledge about meaningful measures and required actions with respect to rebuilding of the chosen area. These observations are shown in a simplified mind map in Figure 7.14.

Based on the results of this soundscape study, several actions were identified for optimizing the general (sound) appraisal of this public space to meet the requirements concluded from the mind map. Derived holistic actions included safety, light, and sound aspects promoting new actions, behaviour, and activities within the reconstructed area. For example, to enhance

Figure 7.14 Simplified mind map of public space Nauener Platz in Berlin. The numbers represent particular relevant locations indicated by residents in soundwalks. The terms represent core categories, which will guide the development of design of the investigated area.

the perceived safety, a light system was installed and all installations, like noise barriers (gabion walls), were not higher than 1.40 m. This led to a feeling of higher security and safety, which also attracted new visitors. Figure 7.16 shows the increased attendance of the place, illustrating its higher attractiveness.

Moreover, audio-islands realized through specific benches equipped with two loudspeakers were installed at the most relevant listening places (Figure 7.15a and b), where by pushing a button people can hear natural sounds like birdsongs or shingle beach. The aim was to introduce sounds that will not mask unwanted sounds, but draw attention away from the road traffic to nature. Thus, sensory overload and overstimulation were avoided. To get an acceptable noise impact at the playground, the gabion wall was installed, which reduced the noise impact by 2.8 dB(A) (Figure 7.15c).

Figure 7.15 Introduced measures to improve soundscape perception of the public space Nauener Platz in Berlin. (a) Bench with loudspeaker system installed. (b) Ringshaped chair with loudspeaker system installed. (c) Gabion wall with a maximum height of 1.4 m.

Figure 7.16 Visitors of rebuilt public space Nauener Platz in Berlin.

7.4.3 Validation Study to Confirm the Taken Measures and Activities

An investigation for the purpose of validation a year after redesign of the public place showed that the introduced actions and arrangement were well accepted (Acloque and Schulte-Fortkamp, 2011) and led to a higher attractiveness of the place and stimulated new activities of the users. This caused an increase of human-made sounds, which was positively perceived and contributed to an increase of felt vibrancy of the place. Of course, the enhanced attraction of the urban place is not only due to an optimized acoustical environment; the redesigned location is perceived globally as more pleasant and inviting, although the major (unwanted) sound source—the road traffic—was not subject to any changes. Balancing between acoustic needs, architectural planning, and the expertise from people living in the area leads to a new concept of the public place.

In summary, mind mapping refers to the identification of relevant categories and their importance regarding the microscopic local context offering solutions for soundscape improvements. The mind mapping concept focuses on human beings and takes into account their individual evaluation and assessment strategies for the derivation of necessary measures. It reflects the location-specific characteristics, like geography, climate, wind, water, people and their behaviour, buildings, and animals, which contribute to the assessment strategies integrated into the respective mind maps.

Although humans obviously evaluate the acoustic environment individually, any decision-making process is consistently embedded in the social–cultural background. Therefore, the social–cultural frame of reference influences the personal evaluation, leading to more or less inter-individual patterns of perception and assessment. Thus, mind mapping based on individuals' way of perceiving their acoustic environment will also refer to the perception related to the society.

7.5 MAPPING NOTICED SOUNDS

Soundscape strongly focuses on the person present in the sonic environment and how this person perceives and understands the sound field. The meaning of the sound within the context was shown to be important in this respect. Meaning is given to the sounds that people notice. Thus, an important step in mapping soundscape is mapping the sounds that users of the space will notice. Whether a sound attracts attention depends on characteristics of the sound, such as changes in time and frequency, often referred to as its saliency (De Coensel et al., 2009). Whether the sound receives attention and gets noticed also depends on the activity of the person. For mapping purposes, the latter can only be included in a general, person-independent way. Grouping of sounds into more complex auditory objects (cars becoming traffic, bird chirps becoming a dawn chorus) is an example of such a common factor that can be included in a model (Oldoni et al., 2013).

7.5.1 Mapping Noticed Sounds Based on Simulation

The temporal fluctuation of the sound level being a strong determinant for noticing sound, calculation methods used for mapping need to include this temporal fluctuation. For traffic sound, the temporal structure of the sound is mainly determined by the movement of individual vehicles. Therefore, traffic sound events typically have durations of several seconds for a car passing at close proximity to several tens of seconds for a train passage or airplane flyover at high altitude. These dynamics can be simulated by

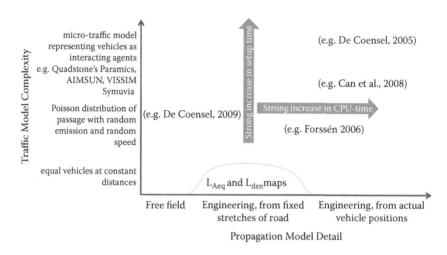

Figure 7.17 Different approaches for modelling the dynamics of road traffic noise.

applying the techniques described in Section 7.1 at 1 s intervals for sound sources moving from time step to time step. Train passages and airplane flight paths can be included explicitly since this information is available at least with a margin of accuracy that is sufficient for the mapping purpose. Road traffic is more stochastic in nature. Traffic microsimulation allows simulating vehicle position and speed for a large number of vehicles, taking into account traffic infrastructure. Alternatively, a Poisson distribution of traffic passages combined with randomized emissions can be used to obtain a first estimate of the traffic noise dynamics. The combination of traffic model and propagation model determine the effort needed (Figure 7.17).

The spectrotemporal dynamics of vocalizations and other nonmechanical sounds is mainly determined by the source, while the propagation path is roughly constant. Thus, it is imperative to have a good idea about the dynamics of these sources in the area under study. In addition, sounds sources such as birds or talking people are mostly distributed over certain areas of the place. Although there are basic data on sound power produced by birds (Brackenbury, 1979), human voices, wind-induced vegetation noise (Bolin, 2009), and water features (Watts et al., 2009) available, it remains difficult to find detailed information on spectrotemporal fluctuation. To preserve ecological validity, it may therefore be advisable to base the mapping of the natural sound component of the sonic environment on measurements within the study area or an equivalent environment (Figure 7.18). If such measurements are not available or cannot be performed, a suitable proxy may be to complement a natural sound spectrum with a $1/f$ modulation of the sound-level envelope since such a statistic is quite common in natural sounds, such as bird chorus or wind-induced vegetation noise (De Coensel et al., 2003).

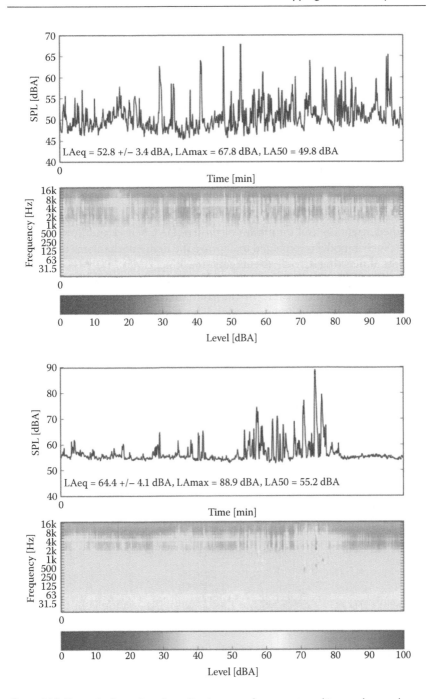

Figure 7.18 Example dynamics of vocalization sounds encountered in an urban park over a 2 min time interval. Left: Children in a playground. Right: Birds near a pond.

Based on the above, time–frequency representations such as the ones shown in Figure 7.18 for all the sounds—positive as well as negative—are present in the area under study, and the sounds that will be noticed can now be estimated for every location. By comparing these spectrograms, masking of one sound by the other can now be estimated. Energetic masking prevents the listener from hearing the masked sound however hard he or she is trying. This type of masking can be calculated based on loudness models such as Zwicker loudness or Moore loudness. However, noticing sound also depends on whether the listener pays attention to the sound. The saliency for each of the sounds considered determines how strongly a sound attracts attention. It can be calculated, for example, using the model proposed in Kayser et al. (2005). Attention switching between sounds that attract attention is further modelled using a competitive, winner-take-all mechanism that accounts for sustained attention and inhibition of return (De Coensel and Botteldooren, 2010). Such a model is expected to capture the main mechanisms underlying perceptual masking as well as energetic masking.

Figure 7.19 shows a case study where both rail and road traffic sound are initially easily noticeable in the triangular open space visible in the upper part of the map (white circle). The planning exercise consisted of

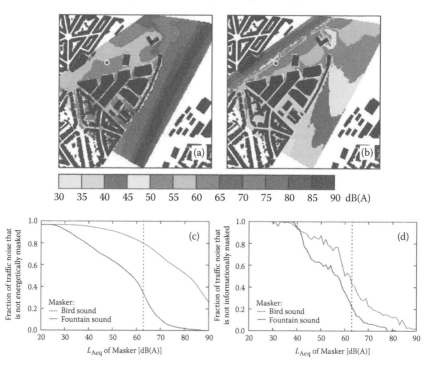

Figure 7.19 (a) Road traffic noise level obtained from microtraffic simulation. (b) Rail traffic noise level. Illustration of energetic (c) and perceptual (d) masking by natural sounds at the location of the dots shown in (a) and (b).

determining the level of natural sound (fountain or bird sound) that would be required to reduce the fraction of the time that traffic sound is noticed by the average visitor of the place, taking into account either energetic or perceptual masking (Figure 7.19c and d). By taking into account modelled spectrotemporal fluctuation of traffic sound and measured spectrotemporal fluctuation of natural sounds, it could be shown in this case that a level of natural sound of 60 to 70 dBA would be needed to obtain a significant reduction in noticing traffic sound.

7.5.2 Mapping Noticed Sounds Based on Measurements

Mapping noticed sounds in existing situations can also be based on measurements. Soundwalks have become a popular way to assess the quality of the urban sound environment. These are usually accompanied by punctual measurements where the group of participants is expected to be quiet. For mapping purposes, it may be more efficient to measure while walking using a sound measurement device equipped with GPS. Such measurement walks have to be conducted with care to avoid the sound of footsteps or clothes dominating the observation. Routes can be preselected to match commonly used paths, and short breaks at places that would be used for that purpose by the daily users of the space can be included. Based on typical duration of noise events, measurements have to be recorded at subsecond intervals, typically 1/8 or 1/10 of a second to allow interpretation afterwards. Based on these sound recordings, sound maps could be calculated showing the sounds that a visitor following this path would most likely notice using the type of models described above. Figure 7.20 shows the probability for noticing natural sounds (birds, water, etc.) on the one hand and human sounds (voices, cries, footsteps, etc.) on the other in a park in Antwerp. For reference, 1 min median noise levels are also shown. In the southern part of the park, high-traffic noise levels prevent natural or human sounds from being noticed. Natural sounds are more likely to be noticed in the more eastern part of the park. This matches expectations, as in this area dense vegetation and natural ponds are found.

Figure 7.20 Example of mapping noticed sounds based on measurements: (a) 1 min median noise level during walking (darker red = higher level), (b) spots where natural sounds are most likely noticed during a warm summer day (white), and (c) spots where human sounds are most likely noticed during a warm summer day (white to black indicating higher probability).

REFERENCES

Acloque, V., Schulte-Fortkamp, B. 2011. Validation of the psychoacoustic infra-structure of a public space in Berlin, based on the concept of soundscape. Presented at DAGA 2011, Düsseldorf, Germany.

Barclay, M., Kang, J., Sharples, S. 2012. Combining noise mapping and ventilation performance for non-domestic buildings in an urban area. *Build. Environ.*, 52, 68–76.

Bolin, K. 2009. Prediction method for wind-induced vegetation noise. *Acta Acust. united Acust.*, 95(4), 607–619.

Boubezari, M., Bento Coelho, J. L. 2012. The soundscape topography, the case study of jardim d'Estrela. Presented at Internoise 2012, New York.

Boubezari, M., Carnuccio, e. t. E., Bento Coelho, J. L. 2011. Soundscape mapping: A predictive approach. Presented at Proceedings of ICSV18 International Congress on Sound and Vibration, Rio de Janeiro, Brazil.

Brackenbury, J. H. 1979. Power capabilities of the avian sound-producing system. *J. Exp. Biol.*, 78, 163–166.

Can, A., Leclercq, L., Lelong, J. 2008. Dynamic estimation of urban traffic noise: Influence of traffic and noise source representations. *Appl. Acoust.*, 69(10), 858–867.

De Coensel, B., Botteldooren, D. 2010. A model of saliency-based auditory atten-tion to environmental sound. Presented at 20th International Congress on Acoustics, ICA 2010, Sydney, Australia, August 23–27.

De Coensel, B., Botteldooren, D., De Muer, T. 2003. 1/f noise in rural and urban soundscapes. *Acta Acust. united Acust.*, 89(2), 287–295.

De Coensel, B., Botteldooren, D., De Muer, T., Berglund, B., Nilsson, M. E., Lercher, P. 2009. A model for the perception of environmental sound based on notice-events. *J. Acoust. Soc. Am.*, 126(2), 656–665.

De Coensel, B., De Muer, T., Yperman, I., Botteldooren, D. 2005. The influence of traffic flow dynamics on urban soundscapes. *Appl. Acoust.*, 66(2), 175–194.

EU. 2002. Directive (2002/49/EC) of the European Parliament and of the Council—Relating to the assessment and management of environmental noise.

Forssén, J., Hornikx, M. 2006. Statistics of A-weighted road traffic noise levels in shielded urban areas. *Acta Acust. united Acust.*, 92(6), 998–1008.

Genuit, K., Schulte-Fortkamp, B., Fiebig, A. 2008. Psychoacoustic mapping within the soundscape approach. Presented at Internoise 2008, Shanghai, China.

GRASS Development Team. 2008. Geographic resources analysis support system (GRASS) software. Michele all'Adige, Italy.

Hao, Y., Kang, J., Krijnders, J. D. 2013. Incidence of green area in context of urban morphology from the viewpoints of both audibility and visibility. *Landsc. Urban Plan.*, accepted.

ISO. 1993. ISO 9613: Attenuation of sound during propagation outdoors. Part 1 (1993): Calculation of the absorption of sound by the atmosphere. Part 2 (1996): General method of calculation. International Organization for Standardization.

Kang, J. 2000. Sound propagation in street canyons: Comparison between diffusely and geometrically reflecting boundaries. *J. Acoust. Soc. Am.*, 107, 1394–1404.

Kang, J. 2001. Sound propagation in interconnected urban streets: A parametric study. *Environ. Plan. B Plan. Design*, 28, 281–294.

Kang, J. 2002a. *Acoustics of Long Spaces: Theory and Design Guidance.* London: Thomas Telford Publishing.

Kang, J. 2002b. Numerical modelling of the sound field in urban streets with diffusely reflecting boundaries. *J. Sound Vib.,* 258, 793–813.

Kang, J. 2005. Numerical modelling of the sound fields in urban squares. *J. Acoust. Soc. Am.,* 117, 3695–3706.

Kang, J. 2007. *Urban Sound Environment.* London: Taylor & Francis.

Kang, J., Chourmouziadou, K., Sakantamis, K., Wang, B., Hao, Y., eds. 2013. *Soundscape of European Cities and Landscapes.* Oxford: EU COST.

Kayser, C., Petkov, C., Lippert, M., Logothetis, N. K. 2005. Mechanisms for allocating auditory attention: An auditory saliency map. *Curr. Biol.,* 15(21), 1943–1947.

Liu, J., Kang, J., Behm, H. 2014a. Birdsong as an element of the urban sound environment: A case study concerning the area of Warnemünde in Germany. *Acta Acust. united Acust.,* 100, 458–466.

Liu, J., Kang, J., Behm, H., Luo, T. 2014b. Effects of landscape on soundscape perception: Soundwalks in city parks. *Landsc. Urban Plan.,* 23, 30–40.

Liu, J., Kang, J., Behm, H., Luo, T. 2014c. Landscape spatial pattern indices and soundscape perception in a multi-functional urban area, Germany. *J. Environ. Eng. Landsc. Manage.,* iFirst, 1–11.

Liu, J., Kang, J., Luo, T., Behm, H., Coppack, T. 2013. Spatiotemporal variability of soundscapes in a multiple functional urban area. *Landsc. Urban Plan.,* 115, 1–9.

Matsinos, Y., Mazaris, A., Papadimitriou, K., Mniestris, A., Hatzigiannidis, G., Maioglou, D., Pantis, J. 2008. Spatio-temporal variability in human and natural sounds in a rural landscape. *Landsc. Ecol.,* 23, 945–959.

Meng, Y., Kang, J., Smyrnova, Y. 2008. Numerical modelling of sound fields with mixed specular and diffuse boundaries using combined ray tracing and radiosity method. In *Proceedings of the Institute of Acoustics (IOA) (UK),* vol. 30, no. 2.

Oldoni, D., De Coensel, B., Boes, M., Rademaker, M., De Baets, B., Van Renterghem, T., Botteldooren, D. 2013. A computational model of auditory attention for use in soundscape research. *J. Acoust. Soc. Am.,* 134(1), 852–861.

Schulte-Fortkamp, B., Fiebig, A. 2006. Soundscape analysis in a residential area: An evaluation combining noise and people's mind. *Acta Acust.,* special issue on soundscapes, 96(6), 875–880.

Schulte-Fortkamp, B., Genuit, K., Fiebig, A. 2008. Developing a public space in Berlin based on the knowledge of new experts. Presented at Internoise 2008, Shanghai, China.

Smyrnova, Y., Kang, J. 2010. Determination of perceptual auditory attributes for the auralization of urban soundscapes. *Noise Control Eng. J.,* 58, 508–523.

Van Renterghem, T., Hornikx, M., Forssen, J., Botteldooren, D. 2013. The potential of building envelope greening to achieve quietness. *Build. Environ.,* 61, 34–44.

Van Renterghem T., Salomons, E., Botteldooren, D. 2006. Parameter study of sound propagation between city canyons with coupled FDTD-PE model. *Appl. Acoust.,* 67(6), 487–510.

Wang, B., Kang, J. 2011. Effects of urban morphology on the traffic noise distribution through noise mapping: A comparative study between UK and China. *Appl. Acoust.*, 72, 556–568.

Watts, G. R., Pheasant, R. J., Horoshenkov, K. V., Ragonesi, L. 2009. Measurement and subjective assessment of water generated sounds. *Acta Acust. united Acust.*, 95(6), 1032–1039.

Yang, W. 2005. An aesthetic approach to the soundscape of urban public open spaces. PhD thesis, School of Architecture, University of Sheffield, Sheffield, UK.

Yang, W., Kang, J. 2005a. Soundscape and sound preferences in urban squares. *J. Urban Design*, 10, 69–88.

Yang, W., Kang, J. 2005b. Acoustic comfort evaluation in urban open public spaces. *Appl. Acoust.*, 66, 211–229.

Yu, L., Kang, J. 2009. Modeling subjective evaluation of soundscape quality in urban open spaces: An artificial neural network approach. *J. Acoust. Soc. Am.*, 126, 1163–1174.

Chapter 8

Approaches to Urban Soundscape Management, Planning, and Design

J. Luis Bento Coelho

Instituto Superior Técnico, University of Lisboa, Lisbon, Portugal

CONTENTS

8.1 INTRODUCTION

Our understanding of the outside world results from a complex cognitive process where all our senses intervene, as described in previous chapters. Coherence between all precepts, together with personal history and culture, in context conveys an overall image of each person's space.

The quality of our environment is perceived from an information processing procedure where all senses and cultural factors play a role, each in a different measure, but all contributing their part. Perceiving the sound around us in the sense that we hear and relate what we hear with the whole reality is inherent to the process of understanding and appraising our space, and thus the world around us (Thibaud, 1998; Kang, 2007).

The economic and social development of the last decades has led to generally higher requirements for quality of life and well-being from people all over the world. The aging of the population may also help explain such higher expectations, since older people bring more experience to the perception process and may be more aware of their rights to a better quality of life. And this trend will continue in the foreseeable future, according to predictions of the worldwide growth of population in urban settings (UN, 2011).

When new projects are envisaged, designed, and implemented, the management of the soundscape becomes a must, in a manner similar to that generally understood for the landscape (Hellstrom, 2003; Kang, 2007; Zhang and Kang, 2007). This is more crucially so in urban environments where the complexity of activities will interfere with the also complex and varied expectations of the urban population. In some places, one will require quiet and tranquillity; in others, one will like a varied soundscape related to one's particular activities; and in yet others, one may even appreciate some mechanical sounds and urban vibrancy that show how dynamic and active the city is.

Architects, engineers, environmental experts, and urban technicians and planners may all be well intended and knowledgeable in their professional roles designing a new project, be it a new urban development, a new transport infrastructure, or a new leisure area, for example, but they frequently overlook its sonic component (Hellstrom, 2003), that is, the sound environment that will be offered to the public space user. This might be a serious issue in the final appraisal of the project, as the soundscape must also be managed and planned, using adequate tools (Zhang and Kang, 2007; Kang and Zhang, 2010), to ensure that acoustic comfort is felt by the listener. Information and criteria regarding the degree of acceptance of the outdoor sound by the end users, basically the citizens, must be incorporated at the drawing board.

Large-scale projects are always subject to environment impact assessment processes where the most relevant aspects are studied. The acoustic environmental studies focus mostly on compliance with the applicable legislation (at different levels), their criteria, and limit values (usually noise limits for different periods of the day or maximum allowed differences). Noise mitigation measures will then be recommended and adopted when the project fails to comply with target (legally binding, recommended, or good practice) noise limits, when applicable quality criteria are not met, or when health issues are too obvious (also usually by following legal or recommended criteria). Even so, such measures at rarely defined at the very early design stages. Experience unfortunately shows that such processes are not able to fully predict the real impacts on the quality of the environment, and that the usual methodologies lack consideration of the human perception mechanisms and their intervening aspects (Lercher and Schulte-Fortkamp, 2003).

A new project is usually analyzed or designed in terms of function (which is right), considering the resulting visual aspect, as one seems to favour visual information, so landscaping is within the scope of most projects, and also some other aspects related to lighting and air quality. It is quite rare, though, that the same consideration is given to the resulting acoustics: to how the sound will work, except for specific-purpose buildings, such as concert or theatre halls or opera houses. Yet, auditive information is a prime input to our mental appreciation process and for the construction of our sense of space (Kang and Zhang, 2010).

Soundscape information needs to be organized, analyzed, and planned for a new project to really succeed. It needs to be managed as part of the whole design. This is not necessarily a complex process requiring high costs. If taken upstream of the conception framework, it may not even involve extra costs. And the results may be very different in terms of the resulting impacts of the project, as the case studies described in Chapter 10 clearly show.

Soundscape planning and design should then be integrated as early as possible in the design stage of a new project.

8.2 OUTDOOR SOUNDSCAPE

Outdoor soundscape embodies complex concepts as stated in the previous chapters. The sonic environment, as it is perceived, understood, and experienced by the user of a space, forms the perceptual construct. The interaction between the user and his or her environment in the place where perception and action occur makes for a dynamic and personal concept.

Soundscape can be a complex structure of perceived sounds, in either urban or nonurban settings, perceived in the specific context (Axelssön et al., 2010). All sound components, of natural or anthropogenic origin, will generally be heard and understood (except where masked), if not just completely or rationally identified, according to their topologies. The degree of appreciation will result from the experience of the user, from the interaction with information from all senses, and from confrontation with his or her expectations in view of the uses of the place.

Noise here is the result of an interpretation of the sound contents (too much energy, awkward frequency spectra or time history, or meaning) or of its context, where the coherence between the different components is found to be unpleasant. Sounds from mechanical sources (such as traffic or building equipment) are usually described as noise. However, they may not necessarily or always be perceived as unpleasant or undesirable, since they might in some cases provide interesting information on urban activities and contribute to the sense of place (Anderson et al., 1983), as described in the following. This then calls for a careful analysis of the situation introduced by a new project regarding the changes that will be introduced in the sonic atmosphere.

The concept may seem too fluid, since it will depend on the place, the users, and their activities, both existing and planned. However, it is the fact that the human listener/perceiver is placed at the centre of the listening process that makes for the complexity, but also for the interest of the study.

The different sound components will be spatially distributed in accordance with the sound wave radiation and propagation laws in free space.

Therefore, the various sounds will be perceived differently, due to the laws of physics, as the user moves in space. Sound reflection, diffraction, or absorption by the land, objects, or materials, and effects of the meteorological conditions will model the whole composition, leading to the richness and variety of the acoustic environment. However, physics will not be enough to define the whole sonic picture. The final soundscape will result from the coherence with all other senses and with the reality of the place and the user (his or her activities).

Would the soundscape in Times Square, New York, be found more interesting if the cars were taken away? Certainly, the noise levels would diminish, but an odd ambient would be felt, perhaps not so interesting since the dynamics associated with that place would not be recognized. The sound here is a resource that is part of the character of the square, not just noise considered a waste (Brown, 2009). Soundscape management calls for much more than controlling noise.

8.3 NOISE CONTROL AND SOUNDSCAPE MANAGEMENT

Architects, engineers, and urban planners are responsible for designing and developing new spaces or redeveloping existing ones, either in cities or in the countryside. Any new project will in principle be designed for function, usually following a multicriteria approach. It should also fit into the surrounding environment and ensure adequate conditions for the users. These conditions will be defined at least by compliance with all health and safety codes, with environmental regulations, and with various other criteria established in applicable national or local ordinances.

A new project or its implementation might have to incorporate environmental impact mitigation measures where and when necessary as required by the fulfilment of the applicable legal obligations and to avoid serious disturbance (where the definition of *serious* is not infrequently left to the whimsical judgement of the design team by lack of proper criteria) of the environment, namely, those regarding noise issues. However, environmental impact assessment studies consider the various environmental components and their interactions, but they frequently fail to bring the human perception to the centre of the analysis. The acoustical studies are usually more concerned with compliance of ordinances and regulations, as well as conservation of previously existing conditions, than with the quality of the sound environment and how it will be perceived by the listeners and contribute to their well-being.

Regarding the acoustic environment, negative impacts usually relate to the increase in sound pressure levels or in population exposed to noise, whereas positive impacts are concluded if less noise or lower exposure values result. Obvious health effects are considered, especially when noise–dose relationships exist. However, effects on the well-being of the citizens and their acoustic comfort in public space are rarely approached. Little consideration is given to the degree of appreciation of the local soundscape and its meaning for the populations affected, either local citizens or visitors.

High sums of money are sometimes spent just to fulfil legal obligations without considering those really involved. A number of situations are known of acoustical barriers designed for protection of noise from railways that had to be either dismantled or not built at all because the population did not want them. On the other hand, the fulfilment of legal noise criteria is not reason enough to conclude for the goodness of the project in the acoustic sense.

It is certainly deemed necessary to reduce sound energy from mechanical origins where they are found to be detrimental to human health and well-being. The effects of environmental noise on health are well known (Lercher, 2003, 2007; EEA, 2010; see Chapters 3 and 5), and the World Health Organization has recommended target values and objective criteria for noise levels in different situations, both indoors and outdoors (WHO, 1999, 2009, 2011). These recommendations have been adopted in most parts of the world in the form of noise limits, design or licensing guidelines, or quality standards for building construction, work, or public places (Bento Coelho, 2007).

Noise control engineers have a crucial role in establishing plans and defining solutions for noise reduction and control. Noise mitigation measures are typically defined at the source, in the propagation path, or at the receiver. The closer to the source, the more efficient the solutions usually are since they will control the direct sound emissions. Usually, these are also the most cost-efficient. However, in complex projects, a blend of solutions applied at the source and on the transmission path may be found to be the most interesting strategy, in both technical and financial terms. Costs can play a very important part in the management of noise issues, and the criteria for adoption of noise abatement measures should always include cost–benefit ratio analysis as a decision criterion.

Acousticians are frequently called at a late stage in the design process when the whole concept has been defined, invalidating some simpler or more obvious acoustic engineering solutions. In some cases, the project can still be adjusted so as to fully comply with limit values or other criteria. Noise mitigation measures will be applied at the source when a transport infrastructure can still change place, even if slightly, but enough to diminish noise at the receiver; when an especially noisy equipment can be replaced by a quieter one; when it can be engineered to have its sound power or sound radiation efficiency diminished; or when an alternative, more favourable noise emission process is selected. In other situations, the project may be able to accommodate proper noise control solutions even if a compromise is reached, such as when operating (traffic or equipment) restrictions are introduced as project conditions.

A new road with high traffic densities may be subject to speed limitations near schools or residential areas without much problem if the speed reduction is not drastic. Or a new shopping mall may see its opening times restricted if the impact on the neighbourhood needs to be controlled. Alternatively, the project may be implemented with noise control solutions introduced in the transmission path, such as noise barriers (either artificial or making use of other constructions, existing or planned), or in limit situations, at the received, when the building façade sound insulation is reinforced, mostly in the glass panes.

These noise control measures are designed basically to reduce excessive sound levels or particular frequency contents of the sound signal that are found disturbing or dominant in the spectral contents, in order to comply with regulations and protect health. This is a necessary stage in every project when sound emissions have to be considered and excessive noise might be an issue.

The work of the noise control engineer, though very important, usually stops here, when compliance with noise limits is achieved. In large projects and installations, noise management programs are drawn, but normally the goal is noise reduction and control aiming at fulfilling legal noise criteria. These are defined considering mean conditions and well-established effects on large populations, thus catering to ensure minimal quality conditions. Therefore, a really successful design may have to go beyond the strictly legal criteria requirements, further ensuring a pleasant experience of the place by its users.

Noise mitigation will diminish sound energy irrespective of the sound contents or context. A noise barrier will attenuate noise from a nearby road, but may also stop the local residents from hearing the bells from a church on the other side of the road, which some people will enjoy listening to. Or the new multiple glass windows will impede them from listening to the birds singing in the garden that they were used to. A number of noise barriers designed for railway projects to reduce the noise received by nearby residents had to be abandoned due to the opposition of those residents who did not care for the "noise" (raising the issue of the classification of noise) and wanted to keep watching and listening to the passing trains.

Reducing sound levels does not necessarily lead to improved quality of life, especially in urban areas. Adequate noise reduction is not always feasible or cost-effective, and more importantly, it will not necessarily lead to improved living conditions and people's satisfaction. Expectations for the sound environment are certainly different for urban and nonurban areas (De Coensel and Botteldooren, 2006), and even inside cities, they depend on the neighbourhoods. Sounds from people talking loud outside or from road traffic may be annoying and understood as noise in a quiet residential or school area, but may be neutral and even welcome in a central city apartment area near restaurants and underground stations, where all urban facilities are available, and its vibrancy was chosen and might be appreciated by the citizen.

A new project, at either a design or assessment stage, must be approached from different angles by considering both the reduction of excessive noise and the maintenance or creation of the feeling of comfort and well-being that people would expect. These aspects are actually part of the same perspective of designing for humans, placing sound as perceived well upstream of the project design and management task and at the centre of the design decisions (Schulte-Fortkamp and Fiebig, 2006). Noise control is directed toward sounds of disturbance and discomfort, some of which will cause serious health effects, as largely reported in the literature, and where sleep disturbance plays a major role (EEA, 2010). The process disregards sounds of preference, which are wanted, desired, expected, and contribute to relaxation and psychological restoration and are essential for the well-being and thus to health and quality of life (Lercher, 2003).

The soundscape approach is a more complete, if a more complex, process, but it leads to much more satisfactory results than classical noise engineering. It requires the cooperation of technical and human/social knowledge; however, that should not be a problem for practitioners when practical guidelines are available and a simple methodology is followed. As the room acoustics expert knows that concepts of speech intelligibility, musical clarity, and others that might be thought to be outside his strict field of technical expertise (the acoustician is not always a musician or a linguistic expert) have to be brought to the optimal room design, so the soundscape analyst or planner has to consider the whole process of perception of a place by the human, where the sound is only a part, though a very important one.

Not just the design framework is different, as described in the next sections of this chapter, but also measurement and prediction techniques must differ from those usually used for noise assessment purposes, as mentioned in the previous chapters.

In fact, common sound pressure level measurements abide by applicable standards (namely, the ISO 1996), where the concept of representative long-term mean value is nuclear. However, people do not stay long in outside public areas. Therefore, the time frame for averaging sound outdoors within the scope of soundscape analysis has to be accordingly short. Sound events and time variations (either seasonal or temporary), which might not be too relevant for the acoustician assessing the outdoor sound field (when measuring overall sound pressure levels), have to be considered with the same degree of importance as sound energy or frequency spectra since they might relate with the rhythms of activities or the uses of the place. The designer is interested here not just in the sound signal characteristics, but also in the meaning and contents, in the context, in coherence with other senses. Therefore, audibility of the different contributing sound components is important, as it is their correlation with the landscape (Pheasant et al., 2008, 2010) and with the different human activities either existing or planned for the place.

Prediction techniques may follow more common lines and procedures as long as these issues are taken into consideration. Soundscape mapping can make use of sound propagation properties and usual mapping software, provided that basic data are acquired, taking perception into account in the assessment procedure. The same is true for other techniques that are used for modelling and planning, such as auralization, which helps to preview and understand future scenarios.

It thus follows that soundscape analysis and design must not be seen as an alternative or even a complement to the classical energy-based noise control engineering. Being a more powerful method and considering human health and well-being in a broader sense, it should be developed in parallel with the classical method that it actually incorporates. Noise control then becomes just one of the procedures available to the practitioner who undertakes an integrated approach that includes soundscape analysis and design.

8.4 SOUNDSCAPE PLANNING AND DESIGN

8.4.1 Managing Soundscapes

Outdoor spaces offer a wide variety of experiences for the different senses of the user. Those experiences will differ depending on the activities and motivations of the user of the place and the user himself or herself (his or her cultural background and other personal aspects) (Brambilla et al., 2006; Easteal et al., 2014). Visually and sonically, for example, a forest or an urban park or a city square will vary in different moments of the day, during the year, for different types of people, very much depending on the activities of the people in such places. Each place, however, bears its own character, comprising the landscape, the soundscape, and the lightscape, with seasonal, geographical, or other variations being an integral part. The soundscape is a major part of the character of every place, and as such, it should be well understood and properly managed.

The management of soundscape is an important stage of the assessment or design of a new project, be it an urban development or a new infrastructure. Even when noise issues are not at stake, because noise sources are either not relevant or nonexistent, the people that will use the place will have a sonic experience that should be as good and enjoyable as possible.

The visual structures in a project are usually managed by building and landscape architects, garden planners, and designers. However, very seldom do soundscape planners integrate the design team. This has been accomplished in some cases by sound artists who introduced sonic elements integrating the design.

Management of soundscape, though being a complex process, is not an intricate or esoteric way of approaching sound outdoors.

The acoustical design of specific indoor spaces has been studied and researched for a long time, and sets of criteria and procedures are well established for room acoustics. Successful theatre and concert halls, auditoria, and classrooms, for example, have to comply with criteria that are well related to the sound perceived by the user, in terms of function.

Outdoor spaces do not seem to have deserved the same attention of sound design according to function. Yet, our degree of discomfort can be felt in many places due to the poor sound character of the place (Yang and Kang, 2005), whereas we find some other places that are found quite pleasant to lead us to stay for rest and change from other, more sonically aggressive urban ambients.

Quiet areas fall in the latter category, even if they are not truly quiet (in strict terms of featuring much less sound pressure levels) sometimes (EEA, 2014). However, in quiet areas the soundscape is able to respond to our needs of a higher-quality environment and a sonic refuge. Nevertheless, other places in central urban areas with traffic may also be found to be sonically interesting since the sounds there can be coherent with the existing dynamics and vibrancy that might be appreciated.

These concepts can be gathered in a structured methodology for design purposes.

8.4.2 Methodologies

Different professionals and practitioners, such as architects, engineers, artists, or urban planners, are involved when outdoor areas are developed or new projects are implemented. In either case, changes in existing soundscape can be expected.

As already mentioned, most acoustical studies of new projects only go as far as assessing the increase of the noise levels, considering the correlation between quantity of sound energy and health effects. This is fine, but not enough, since the quality of the sound and its correlation with well-being and health is also quite relevant, but then completely overlooked. Classical noise control procedures will fall short of the aim of obtaining the best sonic results for the users.

As also explained, new projects gain immensely when a more complete soundscape approach is adopted. An integrated approach where not just noise reduction is considered, but where the soundscape is managed, will put the user at the centre of the project and lead to harmonious solutions and results as much as possible in line with people's expectations.

A soundscape management procedure follows a path that combines simple principles of space planning and perceptual understanding. The methodology proposed by Brown (2012; Brown and Muhar, 2004) for soundscape planning and management is quite clear and straightforward and follows five basic steps: (1) define function or dominant activities of the place, (2) establish unambiguous acoustic objectives where sound contents are made clear, (3) analyze sound contents identifying sounds of preference and wanted and unwanted sounds, (4) acoustically evaluate and assess the sound at the place, and (5) study design options for managing wanted and unwanted sound components so as to achieve the proposed acoustic objectives.

The initial definition of the role of the place in the area and for the listener is crucial since the overall experience (not just aural, but also visual, or light) will depend on the activities and operations that will occur. A similar initial step is taken by the room acoustics designer since the same physical space has to be designed differently for theatre, lectures, or musical performances, for example. Certainly, in acoustic terms, a good concert hall will not be a perfect room for a lecture, and a good conference auditorium may not perform well with an opera. The same applies to open spaces or even to indoor public spaces.

The acoustic objectives will be related to the expected quality of the sound at the place, to the coherence with other senses, and to the information contents of the sound. Table 8.1 presents examples of such objectives (adapted from Brown and Muhar, 2004).

Then it will be important to distinguish which sounds (both existing and future, after intervention) are and will be interesting, wanted, or preferred, and those that will not be wanted or even disturbing. This step is a delicate one since it depends not just on the previous decisions, but also on many conditions, such as time of day or year, weather, and local culture.

Before different or alternative design options are considered or put forward for consideration, all sound component signals must be analyzed in detail in energy, frequency, and time history. The limits of audibility of the different components should be assessed and mapped (see Chapter 7). These components can then be managed, eliminating or reducing unwanted sounds as much as possible, and maintaining, enhancing, or introducing preferred components. This is quite relevant for new transportation projects where a mix of typical pleasant and unpleasant sounds may exist. The overall balance has to be made interesting.

Table 8.1 Examples of Acoustic Objectives

- Moving water or sounds of nature should be the dominant sound heard.
- Only the sounds of nature should be heard.
- A specific sound should be clearly audible over some area.
- Hear mostly (nonmechanical, nonamplified) sounds made by people.
- Not be able to hear the sounds of people.
- Suitable to hear unamplified/amplified speech (or music).
- Acoustic sculpture/installation sounds should be clearly audible.
- Sounds conveying a city's vitality should be the dominant sounds heard.
- Sounds that convey the identity of place should be the dominant sounds heard.

Five points for the process of soundscape design are also proposed by Siebein (2010, 2013): (1) inspiration, (2) planning, (3) conceptual structure, (4) tectonics, and (5) detail. These consider seven elements: (1) identifying the acoustical communities in the soundscape that share common interests, (2) developing taxonomies of the specific acoustic events in the soundscape, (3) mapping the acoustic itineraries of listeners, (4) identifying different spaces impacting on the soundscape of the place, (5) identifying the rhythm of specific events in the soundscape, (6) designing sonic interventions, and (7) implementing solutions.

At the initial inspiration stage, the designer will define the acoustic identity of the project according to location, program, culture, and context. This procedure is usually followed for a theatre or a concert hall acoustical design, using, again, the room acoustics analogy, but usually the acoustic character of the outdoor place or the urban project is not well catered for, especially in urban redevelopment plans where this step is deemed quite important.

At the planning stage, interested acoustical communities must be involved to discuss the goals suggested by inspiration and decide actions and interventions considering the taxonomies of sounds and the different expectations of the users of the place. In urban areas, subareas are found where the goals may differ, depending on aspects like already existing sounds or existing or planned human activities. These differences will be very apparent in larger areas where varied activities or urban topologies can be identified.

The conceptual structure step will organize the space and sound emissions and immissions.

While the tectonics stage defines the whole arrangement of constructions and sound sources that establish the sound distribution and experiences of the listeners, the detail stage "addresses the subtle textures, tones, diffuse, scattered sounds that add coloration to sounds" (Siebein, 2013, 161).

Although these methodologies may seem somewhat conceptually different, they actually follow similar paths. They show a clear structuring of design actions, in the line of common architectural or engineering processes, managing the soundscape at the initial stage of the project design, and considering all aspects of sound composition, contents, meaning, and context.

These methodological procedures can be organized and restructured to form simple guidelines for the practitioner.

8.4.3 Soundscape Design Guidelines

The listener and his sonic interests must be put at the forefront of the design process, with his or her preferences and expectations according to activities, location, and culture. The final objective is that acceptability and identification with the place, together with feelings of comfort, satisfaction, appreciation, and well-being, are achieved.

Table 8.2 Soundscape Design Roadmap

Steps	Criteria/Paths	Techniques
1. Establish the acoustic character of the place	1.1. Define purpose and activities 1.2. Define acoustic objectives according to purpose and activities	• Consider project objectives • Involve stakeholders • Consider listener expectations
2. Plan	2.1. Identify listening places and listening itineraries 2.2. Identify sound sources and sound components 2.3. Identify sound propagation paths 2.4. Identify preferred and unwanted sounds	• Perform soundwalks • Identify time and geographical variations • Measure and characterize sound components • Define topologies of sound components • Involve stakeholders • Build catalogue of preferred sounds

3. Design and optimise	3.1. Manage sound components
	   3.1.1. Diminish unwanted sounds
	   3.1.2. Enhance preferred sounds
	3.2. Identify wanted sounds in context
		• Reduce unwanted sounds (noise control)
		• Mask (psychoacoustics) unwanted sounds
		• Divert attention from unwanted sounds (mind masking)
		• Enhance or introduce preferred sounds in context
		• Involve groups of interest

Three main steps can be identified for the soundscape design at the drawing board, each comprising various paths and criteria, and following different techniques:

1. Identify/define the acoustic character of the place
2. Plan
3. Design and optimize

Table 8.2 summarizes these steps. Of course, the paths and techniques must be duly adopted according to circumstances, but they represent basic guidelines and a practical roadmap.

8.4.3.1 Define the Acoustic Character of the Place

The soundscape planning and design process must start by clearly establishing the acoustic character of the place, in accordance with the designated purpose and the planned activities.

The soundscape is part of the character of the place, as are the visual forms, materials, lights, odours, and people using it. The local history, rhythms, and culture are also a part. This step encompasses the consideration of uses and activities of the space and the objective acoustical goals.

An urban park where one can rest, jog, socialize with friends, and contemplate nature will have an acoustic character (one would expect some peacefulness and tranquillity and would appreciate natural sounds such as those from animals, wind in trees, or moving water) different from another park where one will go for a handicraft or antique fair, or from a big children playground, where more lively sounds would fit the place and its purpose. And the characters of these parks will be different from an outside shopping area or restaurant precinct, for example.

Then, "What is the place designed for?" and "Who will use the place?" are the first questions at the drawing table. "What soundscape will those users expect or find attractive for the designed activities?" will be the next one.

8.4.3.2 Planning

A planning step will then start by carefully separating and identifying areas of listening, users' itineraries, sound sources, sound components, sonic interests, and context. A number of practical procedures are necessary to identify, classify, and characterize existing and future sounds. Soundwalks at different times of the day will identify/designate the existing sound components in the various itineraries of the listeners. The dominant sounds, the different rhythms, and the time and geographical variations must also be found.

The sound component topologies should be well defined by evaluating their limits of audibility in the whole area of interest. The sounds must be acoustically measured in the different places under consideration. The acoustic measurements must consider contributing acoustic events and short-term rhythms. Common environmental acoustics long-term averaging schemes will not be appropriate, and the usual environmental noise indicators must not be used.

Of course, the environmental noise indicators L_{den} and L_n (used in the European space, especially for strategic noise mapping purposes, as recommended by European Directive 2002/49/EC) may be useful since they can provide basic quantitative information. L_d and L_e will be a lot more interesting than L_{den} alone, since this is a composed calculated indicator with little meaning in terms of perceived sound. Moreover, it should be noted that these are long-term values and, for example, L_d will refer to the whole-day period, usually defined as a 12-hour period, where many sonic changes may occur. Short-term sound pressure levels, statistical and sound event indicators, and narrow-band indexes will be more adequate for a finer analysis.

The various sounds' topologies can be mapped to establish their areas of influence, thus differentiating the sound composition in the different areas of interest (Boubezari, 2011; Boubezari and Bento Coelho, 2012). The same procedure should be followed for the future sounds resulting from the new project or those that might be introduced or expected. These techniques can help the designer by clearly establishing planning options. A catalogue of existing and future sounds must be organized, and each one should be classified (sound of preference, unwanted sound, or neutral) according to the character of the place (and possible subarea) and expectations. The involvement of the stakeholders, especially the listeners, is important and will help the designer in the building of the sound catalogue or sound identity map (where the different sounds are distributed over the geographical area of interest).

It is important at this stage to clearly distinguish the sounds of preference and the sounds of discomfort or unwanted sounds. It must be stressed that this classification is not an absolute one. Although usually mechanical sounds are considered unwanted, in many circumstances they are welcome or just ignored. It depends on the context of place, the human activities, or the overall sound composition. Church bells are usually interesting if not too loud or too frequent to be found disturbing. They provide a sense of place and a cultural atmosphere. Traffic sounds, especially from conventional trains or trams, can be found to contribute to the sense of urban liveliness in vibrant areas, though they might not be desired in residential areas. People usually prefer natural or human sounds. Sounds from moving water, wind, birds, or people talking are usually wanted sounds. However, in some cases, dogs barking will be irritating and loud conversations may be intrusive and unpleasant, especially at late hours of the evening and at night.

The catalogue of sound components and their geographical areas of interest can then be correlated with the factors set up earlier: (1) general function of the places under analysis; (2) existing and planned human activities; (3) other senses in the same place, such as landscape, light, and smells; and (4) preferred sounds in the place.

8.4.3.3 Design Actions

In the final design stage, options for soundscape management will be considered and discussed with the acoustic groups of interests, namely, stakeholders (e.g., residents, citizen groups, transport authorities), planning technicians (architects, engineers, urban planners, consultants involved), and decision makers (e.g., local authorities).

It may be interesting to consider different scenarios and to use prediction and simulation methods using all available advanced technological tools, such as predictive mapping or three-dimensional visualization and auralization techniques. These are useful for the experts in their design procedures, but most of all for the nontechnically minded people (citizens, decision makers) who will be able to have a preview of the results, which is especially interesting for the end users.

Noise control measures and strategies should be used when and where possible to reduce or eliminate unwanted sounds, especially from traffic or other mechanical origin, especially if they are too high. This is an important step in many urban development plans where changes in mobility, by either introducing or encouraging soft modes of transportation or adopting traffic limitations, can provide reasonable noise reductions.

Masking techniques may also be adopted by making use of the psycho-acoustic masking phenomena. This can be achieved by enhancing or introducing sounds of preference that will mask unwanted sound components. If new sounds are introduced, either naturally or artificially (by electro-acoustic means), care should be taken so that they fit the character of the place as initially established.

Mental masking can also be used where the attention of the listener is diverted to other sounds that are found more pleasant, as mentioned in Chapters 4 and 7. This can be achieved effectively in urban parks, where natural or human sounds can cause the sound from the surrounding road traffic to be virtually ignored by the park users if they are immersed in intense activities or hear competing (in terms of sound pressure levels) sounds.

This stage will manage the soundscape composition by designing the distribution of sound sources and their audibility, the physical structure of the listening places, and by introducing all details that will define the final soundscape. Care must be taken to distinguish all subareas where different listening characters will be established. In more rigid situations, such as redevelopment of historical city areas, the differences may be slight, but they might be crucial if variety and differentiation are used as assets.

Acoustic studies of new urban projects or redevelopments should then follow the above simple steps. Note that the common noise control engineering role is included. However, it is extended to cater for the character of the place, both existing and future, properly managing the soundscape to ensure an improved quality of the place as perceived by the users. Environmental engineers can adopt these guidelines to assess and manage impacts on the acoustic environment, minimizing negative impacts and, most importantly, enhancing and promoting positive ones.

8.5 CONCLUSIONS

Soundscape is an inherent part of the character of an urban place used by local citizens in their daily routine and by visitors alike. It impacts directly on the feeling of well-being, providing an indication of quality of life and determining the accomplishment of the place for the user. Its management is thus as important as its visual landscape, which is usually ensured by the architectural design team in any new project, though that is not usually the case where the soundscape is concerned.

Most transport infrastructures, urban development, or rehabilitation projects deal with sound issues when approaching the noise problems, either existing or brought up by the new project, mostly due to applicable regulations with limit values or noise increase criteria. Noise abatement measures are duly adopted and implemented when noise limits are not met. It is, however, rare that enough attention is dedicated to understanding, planning, and designing the existing or future soundscape associated with the project. And yet, that is so important in the sense that failure in doing so may determine the failure of the whole project and its acceptance by the population.

The acoustic comfort of public places is determined by numerous facts related to the prevalence or just presence of sounds of preference, coherence of sounds with other senses, especially visual, and correlation with the users' activities and expectations. And all these can and need to be managed in a well-designed project to build sustainable and pleasant cities.

Soundscape planning and design should follow simple but solid methodologies that account for all major aspects that determine our perception of the outdoor sound environment. Three major steps, each one comprising various procedures and tasks, can be considered for action: (1) establishment of the character of the place, (2) preparation and planning, and (3) design and optimization.

The character of the place must be well defined in accordance to its geography and topology, the existing and future sound sources, human activities, and expected uses, even if varied in space (when the place is large) and time (differences during the day or throughout the year may arise). Clear acoustic objectives must also be set up partly related to the other factors and responsible for contributing to the sense of place.

In the planning stage, all subareas and sound components are identified and catalogued in space regarding their acceptance as wanted or unwanted according to the established character of the place. Quantification will also be needed so that audibility limits can be assessed and planned.

Design actions will then be set up for reducing unwanted sounds, by noise abatement and control methods, and for enhancing or introducing wanted sounds in context. Psychoacoustic or mind masking techniques can be followed with the final goal of providing a pleasant acoustic image of the place.

Involvement of the groups of interest, stakeholders (especially residents), transport authorities, and urban technicians will help to define harmonious solutions that can be accepted by all to create an urban soundscape that can be enjoyed.

REFERENCES

Anderson, L. M., Mulligan, B. E., Goodman, L. S., and Regen, H. Z. (1983). Effects of sounds on preferences for outdoor settings. *Environ. Behav.* 15(5), 539–566.

Axelssön, Ö., Nilsson, M. E., and Berglund, B. (2010). A principal components model of soundscape perception. *J. Acoust. Soc. Am.* 128(5), 2836–2846.

Bento Coelho, J. L. (2007). Community noise ordinances. In *Handbook of Noise and Vibration Control*, ed. M. J. Crocker, 1525–1532. Hoboken, NJ: John Wiley & Sons.

Brambilla, G., Maffei, L., De Gregorio, L., and Masullo, M. (2006). Soundscape in the old town of Naples: Signs of cultural identity. *J. Acoust. Soc. Am.* 120, 3237.

Boubezari, M., and Bento Coelho, J. L. (2012). The soundscape topography: The case study of Jardim d'Estrela. Presented at Proceedings of Internoise 2012, New York, August.

Boubezari, M., Carnuccio, E., and Bento Coelho, J. L. (2011). Soundscape mapping: A predictive approach. Presented at Proceedings of ICSV18 International Congress on Sound and Vibration, Rio de Janeiro, Brazil.

Brown, A. L. (2009). The acoustic environment as resource, and masking, as key concepts in soundscape discourse and analysis. Presented at Proceedings of Euronoise 2009, Edinburgh, Scotland.

Brown, A. L. (2012). A review of progress in soundscapes and an approach to soundscape planning. *Int. J. Acoust. Vib.* 17(2), 73–81.

Brown, A. L., and Muhar, A. (2004). An approach to the acoustic design of outdoor space. *J. Environ. Plan. Manage.* 47(6), 827–842.

De Coensel, B., and Botteldooren, D. (2006). The quiet rural soundscape and how to characterize it. *Acta Acust. united Acust.* 92(6), 887–897.

Easteal, M., Bannister, S., Kang, J., Aletta, F., Lavia, L., and Witchel H. (2014). Urban sound planning in Brighton and Hove. Presented at Proceedings of Forum Acusticum 2014, Krakow, Poland.

European Commission. (2002). Directive 2002/49/EC of the European Parliament and of the Council of 25 June 2002 relating to the assessment and management of environmental noise. *Off. J. Eur. Comm.* 45, 12–25.

European Environmental Agency (EEA). (2010). Good practice guide on noise exposure and potential health effects. EEA Technical Report 11/2010.

European Environmental Agency (EEA). (2014). Good practice guide on quiet areas. EEA Technical Report 04/2014.

Hellstrom, B. (2003). *Noise Design, Architectural Modeling and the Aesthetics of Urban Acoustic Space*. Bo Ejeby Forlag, Sweden: School of Architecture, Royal Institute of Technology, KTM.

ISO 1996: Acoustics—Description, measurement and assessment of environmental noise—Part 1: Basic quantities and assessment procedures, 2003; Part 2: Determination of environmental noise levels, 2007. International Organization for Standardization.

Kang, J. (2007). *Urban Sound Environment*. London: Taylor & Francis.

Kang, J., and Zhang, M. (2010). Semantic differential analysis of the soundscape in urban open public spaces. *Build. Environ.* 45(1), 150–157.

Lercher, P. (2003). Which health outcomes should be measured in health related environmental quality of life studies? *Landsc. Urban Plan.* 65, 63–72.

Lercher, P. (2007). Environmental noise: A contextual public health perspective. In *Noise and Its Effects*, ed. L. Luxon and D. Prasher, 345–377. London: Wiley.

Lercher, P., and Schulte-Fortkamp, B. (2003). The relevance of soundscape research to the assessment of noise annoyance at the community level. Presented at Proceedings of the 8th International Congress on Noise as a Public Health Problem, Rotterdam, The Netherlands.

Pheasant, R., Horoshenkov, K., and Watts, G. (2008). The acoustic and visual factors influencing the construction of tranquil space in urban and rural environments tranquil spaces-quiet places. *J. Acoust. Soc. Am.* 123, 1446–1457.

Pheasant, R., et al. (2010). The importance of audio-visual interaction in the construction of tranquil space. *J. Environ. Psychol.* 30, 501–509.

Schulte-Fortkamp, B., and Fiebig, A. (2006). Soundscape analysis in a residential area: An evaluation combining noise and people's mind. *Acta Acust. united Acust.*, 96.

Siebein, G. W. (2010). Essential soundscape concepts for architects and urban planners. Presented at Proceedings of Designing Soundscape for Sustainable Urban Development Conference, Stockholm, Sweden, 26–30.

Siebein, G. W. (2013). Creating and designing soundscapes. In *Soundscape of European Cities and Landscapes*, 158–162. Oxford: European Science Foundation (COST).

Thibaud, J.-P. (1998). Comment Observer une Ambiance? In *Ambiances architecturales et urbaines, Les Cahiers de la Recherche Architecturale.* N.º 42/43. Marseille: Ed. Parenthèses.

United Nations (UN). (2011). *World Population Prospects: The 2010 Revision.* New York: Department of Economic and Social Affairs, Population Division.

World Health Organization (WHO). (1999). *Guidelines for Community Noise,* ed. B. Berglung, T. Lindvall, and D. H. Schwela. WHO report.

World Health Organization (WHO). (2009). *Night Noise Guidelines (NNGL) for Europe.* WHO report.

World Health Organization (WHO). (2011). *Burden of Disease from Environmental Noise: Quantification of Healthy Life Years Lost in Europe.* WHO Regional Office for Europe.

Yang, W., and Kang, J. (2005). Acoustic comfort evaluation in urban open public spaces. *Appl. Acoust.* 66, 211–229.

Zhang, M., and Kang, J. (2007). Towards the evaluation, description and creation of soundscapes in urban open areas. *Environ. Plan. B Plan. Design* 34, 68–86.

Chapter 9

Soundscape as Part of the Cultural Heritage

Luigi Maffei,[1] Giovanni Brambilla,[2] and Maria Di Gabriele[3]

[1]Department of Architecture and Industrial Design,
Second University of Naples, Aversa, Italy

[2]CNR-Institute of Acoustics and Sensors "O.M. Corbino," Rome, Italy

[3]Department of Architecture and Industrial Design,
Second University of Naples, Aversa, Italy

CONTENTS

9.1 INTRODUCTION

Imagine yourself in Istanbul, a city included in the United Nations Educational, Scientific and Cultural Organization (UNESCO) World Heritage List. As you jog along the Bosphorus every day with your iPod and headset on, describe this experience to your friends. Although you will probably talk only about the vision of the Palace of Topkapi in the background, with an underlying choppy sea, of seagulls in the sky, of fishermen on the quay, and of ships traversing the strait, your friends will enjoy this experience, combining in their imagination vision and sound of the described scenario: the splash of the sea on the quay, the seagull's call, the voices of the fishermen, and the rumble of the ships' engines (Figure 9.1).

Image yourself now visiting Trevi Fountain in the historic centre of Rome, also included in the World Heritage List (Figure 9.2). You have chosen for your visit a classical venue in which history, culture, and charm combine. You know all about its architectural style, the building materials, the architects involved, and the restoration events in the last centuries, and it is also a sunny day, but despite all of this, during your visit you feel uncomfortable. Suddenly, you become aware that the music from a loudspeaker in

Figure 9.1 A view of Istanbul.

Figure 9.2 A view of Trevi Fountain in Rome. (From Brambilla G., Maffei L., *Noise Control Eng. J.*, 58(5), 532–539, 2010.)

front of a shop covers the expected sound of the fountain with its moving and splashing water.

These two examples point out how human beings integrate, in real or imaginary situations, the sensorial stimuli that wrap them up (Guillén and López Barrio, 2007; Haverkamp, 2007). Buildings, panorama, and in general all cultural and natural heritage cannot be described, appreciated, and consequently, valourized using a monosensorial component analysis essentially based on vision.

When the feature of hearing in a scenario, now, as in the past, is important, and in some cases predominant, in the positive perception of the population, then the soundscape of that scenario should be considered an element to preserve and valourize, just like the other features.

9.2 MULTISENSORIAL APPRECIATION
OF CULTURAL HERITAGE

In ancient times the relative importance among the population of the sense of hearing (in old Greek, *akoé*) with respect to the sense of vision (in old Greek, *opsis*) and vice versa had alternating moments. Let us think to the period in which laws were orally transmitted, or the following period, in which script laws were preferred because of their more trustable feature.

In the last century, instead the technology world highlighted the deep relation between *hearing and vision*, trying to merge these two sensations as much as possible. Considering mass communication, the television surpassed the radio and the newspapers in terms of popularity; as interpersonal communication, there are continuous technology attempts to substitute the classic telephone with more sensorial devices: smartphones are examples of this trend.

Other disciplines, however, show a different approach. Architecture is concerned with the design, arrangement, and manipulation of a space to make it suitable for daily life activities. To communicate the artistic, social, and emotional context of a space, architects almost exclusively consider the visual aspects of a built environment and often underestimate the acoustic aspects (Brambilla et al., 2013). These, besides the aural experience of an environment and the expectations and preferences of the population (residents, tourists), are critically important to the social and emotional well-being (Maffei, 2008).

Architects are often also involved in the restoration of small or large parts of historical sites (squares, streets, façades, buildings) that can constitute the cultural heritage of a place. All restoration projects consider the prior knowledge of several factors, such as the past urban morphology, the original architectural elements and materials, and the transformation over the time; no interest is focused on the sonic environment as it is perceived now or was in the past. The output of these projects is the restoration of the tangible elements of the cultural heritage, but in many cases, discarding the intangible elements, like the soundscape and its composition (keynote sounds, sound signals, sound objects, sound events, soundmarks as described by [Schafer, 1977]); the results of the intervention are not fully appreciated by the population.

In the past, this limited approach was partially justified by the policy of UNESCO, which, with its international treaties, focused all actions on protecting solely the tangible expression of culture (UNESCO, 1954, 1972, 2001).

However, with the Convention for the Safeguarding of the Intangible Cultural Heritage (ICH) (UNESCO, 2003), the close connection between tangible and intangible cultural heritage was recognized by UNESCO and the concept of cultural heritage expanded. The aim of this convention is now the safeguarding of "practices, representations, expressions, knowledge, skills" and "instruments, objects, artifacts and cultural spaces associated therewith, that communities, groups and, in some cases, individuals recognize as part of their cultural heritage."

The recognition of the ICH involves the recovery of the "living" expressions of the culture, orally transmitted from generation to generation and constantly reproposed by the communities. They play a key role in providing the community with a sense of cultural identity, as well as diversity and uniqueness, in a historical period that is profoundly characterized by the process of globalization.

A change in the conservation policy is then required, and it is under development.

The conservation of the sound of church bells, of the reverberant atmosphere inside a church, of the quietness sense in a cloister, of the voices of the shopkeepers in a crowded historical market, of the rattle of a old tram, can be as attractive for the tourists and as important for the local and global community as the restoration of the bell tower, of the paintings of the church, of the plasters of the cloister, and of the façades of the buildings along the street that hosts the historical market or is crossed by the old tram.

9.3 SOUNDSCAPE AS INTANGIBLE TRADEMARK: RECOGNITION TECHNIQUES AND APPLICATIONS

It is undeniable that the recognition of tangible cultural heritage is based on consolidated methodologies, with the main involvement of historians, architects, and archaeologists. On the other hand, the intangible cultural heritage can be seen as a set of knowledge, skills, experiences, and sensations whose recognition needs the involvement of a wider group of disciplines that range from physics to social and human studies. Another peculiarity of the intangible cultural heritage that makes its recognition harder is its time variance, that is, the possibility that features of the intangible element change over time without losing, however, their intrinsic characteristics as recognized and felt by the population. This can be particularly true for the soundscape of a place or of an event that is strongly linked to the social and cultural heritage of the community (Brambilla and Maffei, 2010; Brambilla et al., 2007a, 2007b; Raichel et al., 2004) and changes over the years as the environment, society, and culture change (Brambilla and Maffei, 2011).

Dealing with this aspect an appropriate approach would be a methodological triangulation (Yeasmin and Ferdousour Rahman, 2012; Hussein, 2009; Botteldooren et al., 2013; Lercher and Schulte-Fortkamp, 2013) based on the analysis and criteria regarding the three components of the scenario, physical, historical, and social (PHS), can be a valid support to recognize

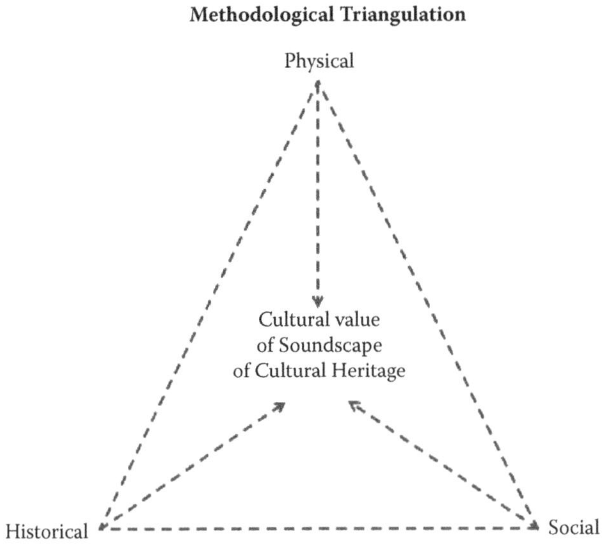

Methodological Triangulation

Figure 9.3 A scheme of the methodological triangulation.

whether the cultural value of the soundscape of a place or event comes out or not (Figure 9.3).

The physical analysis can require on-site acoustic measurements or virtual acoustic reconstructions. The acoustic measurements, performed with several techniques, such as soundwalks, fixed-position measurements, or mono- or biaural recordings, and with several types of equipment, have the scope to identify and classify the sound sources and characterize the temporal and tonal structure of the sound-searching distinction elements. At this stage, measurements and analysis of other physical parameters (vision, odour, microclimate) should be performed because they can influence the human response to the sound stimuli in a multisensorial way. Then the virtual acoustic techniques enable us to reconstruct lost or future audio scenarios (Drettakis et al., 2007; Tsingos et al., 2004; Funkhouser et al., 1999; McDonald et al., 2009; Vorländer, 2008).

The aim of historical analysis is to investigate, mainly through oral information and written documents, but also multimedia sources and meetings with local experts, the evolution across the time of elements representing potential sound sources that could have direct or indirect influence on the soundscape of the place or event. Among these elements, the carefree habits of the population, the crafts, the religious activities, and the transportation system can be relevant.

The main aim of the social analysis is to verify the role of the soundscape in the global perception of the place or event by local and not local communities. During this stage, population expectations, annoyance assessment, cognitive tasks under stimuli, and their correlation with physical stimuli are investigated. This can be done by surveys performed by meetings with expert and nonexpert populations and by laboratory tests. The recent virtual reality (VR) facilities (immersive or not) (Basturk et al., 2010, 2011a, 2011b, 2012), by which past, present, and future scenarios can be reproduced and experienced, are a valuable support to the social analysis, as the reaction of a sample of population to modification across the time of the asset of audio–video stimuli can be investigated and analyzed.

The above description of the PHS methodological triangulation, although conceptually valid, lacks operative details, as the records of tangible and intangible cultural heritage are very wide ranging and generalization is rather difficult. Operative details are described in the following paragraphs through the application of PHS triangulation to three different case studies: a historic city centre, one of the most known archaeological areas in the world, and a folk festival. All these places and events are included in the World Heritage List.

9.4 HISTORIC CENTRE OF NAPLES

According to the description reported by the UNESCO World Heritage Centre (UNESCO, 1995), Naples is one of the most ancient cities in Europe, whose contemporary urban fabric preserves the elements of its long and eventful history. Its street pattern, its wealth of historic buildings from many periods, and its setting on the Bay of Naples give it an outstanding universal value without parallel, and one that has had a profound influence in many parts of Europe and beyond. This description, however, does not take into account other aspects more strictly connected to the intangible cultural value of the site and to the multisensorial perception of the residents or tourists. Several studies (Brambilla and Maffei, 2010; Brambilla et al., 2007b; Di Gabriele et al., 2010) were performed to evaluate the urban soundscape and match it with the cultural value of the site.

The physical analysis was performed along the main roads (*decumani*) and the narrow streets that constitute the ancient urban grid, in the most attractive squares along the paths and inside large cloisters, with their colonnades and gardens that are daily frequented by local residents and tourists (Figure 9.4).

The most relevant sound and psychoacoustic descriptors (L_{Aeq}, statistical levels, loudness, spectrum centre of gravity G) were determined at fixed positions or during soundwalks, which better represent the behaviour of pedestrians that experience the city in a dynamic sequence rather than in fixed positions. As much data as possible were acquired with binaural recordings. During the measurements, types and characteristics of the sound sources were identified. Acoustic data were reported in several ways. For instance, Figure 9.5 shows the variation in time and space of the equivalent sound

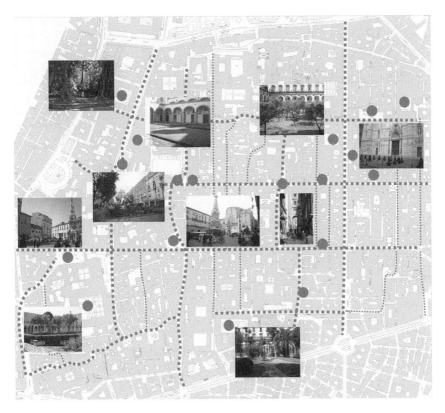

Figure 9.4 Map of the historic centre of Naples. (From Maffei L., Soundscape approach in urban renewal: Parks and areas of acoustic quality, in *Proceedings of DAGA 2010, 36th Jahrestagung fur Akustik*, Berlin, 2010, pp. 51–52.)

levels and loudness levels during a walk: for the observed lowest levels, the corresponding locations are also shown.

Besides sound, other environmental features can influence the emotional reaction in historic districts: visual contamination, spatial perception, light (natural and artificial), microclimate, odours, and architectural degradation. Each of these features can be quantified with numerical parameters, and their variation can be represented along a path, as in Figure 9.6.

The combination of this kind of information leads to sensorial maps in which the experience of a pedestrian moving in the area is expressed in terms of variation of physical stimuli.

A group of historians, planners, and acousticians was engaged in order to examine in depth the past social, commercial, and craft activities and population lifestyles in the area. The study enabled us to hypothesize the location of specific sound sources (craftsmen working with their hand tools, shopkeepers shouting the features of their merchandise, bells, transportation

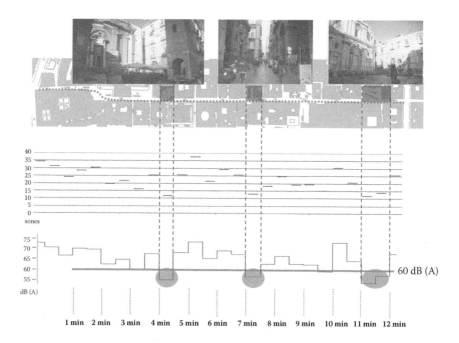

Figure 9.5 Variation of the equivalent sound pressure level and loudness along a walk. (From Maffei L., Soundscape approach in urban renewal: Parks and areas of acoustic quality, in *Proceedings of DAGA 2010, 36th Jahrestagung fur Akustik,* Berlin, 2010, pp. 51–52.)

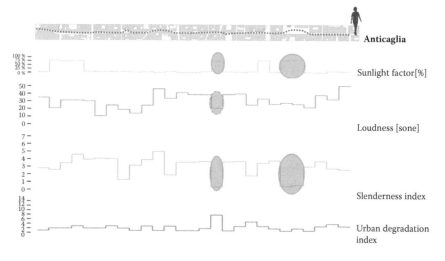

Figure 9.6 Variation of physical features experienced during a walk. (From Di Gabriele M., et al., Urban noise mapping based on emotional dimensions, presented at Proceedings of Euroregio 2010, Congress on Sound and Vibration, Ljubljana, Slovenia, 2010, paper 075.)

Figure 9.7 Street markets in Naples in the past and nowadays. (From Brambilla G., Maffei L., *Noise Control Eng. J.*, 58(5), 532–539, 2010.)

Figure 9.8 A closed cloister in Naples.

systems for people and goods, artists, etc.) and compare these assumptions with the actual situation. It was possible to identify

1. Areas in which, although the people's lifestyle and some details were modified, the actual soundscape can be assumed to be similar to the one in the past. Streets with popular markets are an example (Figure 9.7), as well as closed cloisters considered quiet areas, now, as in the past (Figure 9.8).
2. Areas in which the presence of a specific sound source has altered the original soundscape, which is, however, still present in the background. An example is narrow streets with craftsmen's activities and pedestrians where motorcycle pass-bys are allowed (Figure 9.9).

Figure 9.9 A narrow street in the historic centre of Naples.

3. Areas that had a radical urban and social transformation and in which all past sound sources disappeared and were replaced by new ones. Examples are roads where craftsmen's laboratories have been transformed into shops identifying a commercial area or the transportation urban plan allowing car pass-bys and the presence of parking areas.

The urban space ought to be shaped by the concept of mixophilia, to favour and encourage the possibility of living peacefully and happily with difference, and taking advantage of the variety of stimuli.

From the maps obtained in the physical analysis and historical analysis, it is possible to identify sites where a strong variation of the sound stimuli can be experienced by the pedestrian, enforcing the idea of urban mixophilia and, at the same time, where the soundscape has an invariant character respect to the past.

However, to quantify the extent to which, in these sites, the soundscape could be part of the cultural heritage and how the population is involved in this decision process, a social analysis was needed. This was performed by a laboratory test on a sample of the local population and with on-site interviews. The aim was to investigate how the soundscape of the site can contribute to the recognition of the site itself. For this purpose, binaural recordings and photos of the sites were taken separately and then combined randomly. The subjects participating in the laboratory test had to report on a scale from 1 (none) to 5 (very good) the relevance of the sound with the sonic environment of the site they expected to hear based on the photo of the site.

In another test, the subjects were asked to express on a 7-point bipolar scale judgement on several scenarios of the same site. This laboratory test was aimed at analyzing how the site, in its original visual and audio scenario, is perceived by the subject as an important and essential part of the cultural heritage of the city in which any modification should be carefully examined. The examined scenarios were prepared by merging, through vision and auralization techniques, new design elements and, when possible, new functions of the site with the corresponding new audio signals. The subjects could take part in the experiment using an immersive virtual reality (IVR) system in which the scenarios were reproduced. The IVR technology allows the simulation of an artificial world on a 1:1 scale that can give the observer a sense of being (presence) in the environment and of interaction with it in real time.

Other important information was taken by the survey conducted directly on site with specific interviews of residents and tourists. It was found that some sites are frequented more for their soundscape and the sensation they evoke, rather than for their architectural values. This is the case of the numerous cloisters seen by the population as oases of calmness in the middle of the city or some streets populated by young people and singers, in which the sensation of vibrancy is predominant.

The application of the PHS methodological triangulation led to the identification of sites in the historic centre of Naples where the soundscape is an intrinsic value of the cultural heritage and where all actions should be oriented toward its preservation and valorisation.

9.5 ARCHAEOLOGICAL AREA OF POMPEII

The archaeological areas of Pompeii, Herculaneum, and Torre Annunziata have been included in the World Heritage List considering that "the impressive remains of the towns of Pompeii and Herculaneum and their associated villas, buried by the eruption of Vesuvius in AD 79, provide a complete and vivid picture of society and daily life at a specific moment in the past that is without parallel anywhere in the world" (UNESCO, 1997).

This motivation is very impressive as, besides the tangible cultural value of the residential and public buildings, façades, theatres, and wall paintings, through the words "vivid picture of society and daily life," the cultural value of the intangible elements of the sites is emphasized and, consequently, should be clearly perceived during a visit. However, this is only partially true. Nowadays, walking along the streets of Pompeii, the 2 million people that visit the site each year can see the urban structure and, entering into the buildings, see the rooms and imagine people living inside. They can touch stones and walk on the original pavement, but they are not able to hear the soundscape of 20 centuries ago or smell the flavours used by the local population. It is like watching a movie in which video and audio are not synchronized, or even worse, they are not congruent. This is also the impression that the tourist receives visiting a nearby virtual reality museum in which the vision of the ancient city is guaranteed by

Figure 9.10 Map of the archaeological area of Pompeii. (From Brambilla G., et al., Soundscape in the archeological area of Pompei, presented at Proceedings of 19th International Congress on Acoustics (ICA), Madrid, Spain, 2007, paper ENV-10-002.)

sophisticated three-dimensional reconstructions, but the audio is only a musical score.

The archaeological area of Pompeii (Figure 9.10) covers about 0.7 km², and it is bounded by walls with eight gates. The area is surrounded on the east and west sides by a railway. The south and west sides of the area are also exposed to road traffic, with average flow of 100 vehicles per hour, most of which are tourist buses. The area is also crossed by the air route approaching the close airport of Naples.

Thus, the physical analysis was oriented to the evaluation of the actual soundscape inside the area and to determine the influence of the external sound sources (Brambilla et al., 2007a; Maffei et al., 2012). Soundwalks along the main paths and the less frequented paths of the area were performed to collect binaural recordings. Additional sound measurements were taken in fixed positions where tourists generally stand, receiving instructions by the tour guides. The analysis showed that the main sound sources, not homogeneously distributed inside the whole area, were amplified voices of tourist guides, people shouting, natural sounds (bird twittering, dog barking, wind), car and train passing, and aircraft overflies. Figure 9.11 reports a map of the distribution of the main sources concurring to the sonic environment, and Figure 9.12 shows the corresponding equivalent sound-level map.

It is evident that the soundscape of large parts of the area is influenced by technological sound sources, and maximum sound levels are observed nearby important public buildings where tourists and their guides stand.

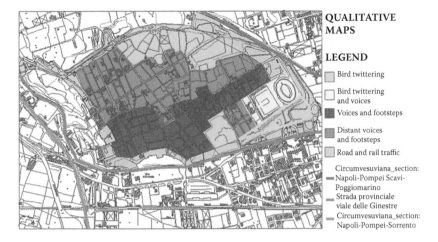

Figure 9.11 Distribution of main sound sources. (From Maffei L., et al., Valutazione del paesaggio sonoro dell'area archeologica di Pompei, in *Atlante di Pompei*, ed. C. Gambardella, 321–330, Napoli: La Scuola di Pitagora Editrice, 2012.)

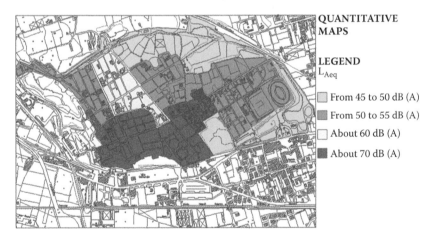

Figure 9.12 Sound-level map. (From Maffei L., et al., Valutazione del paesaggio sonoro dell'area archeologica di Pompei, in *Atlante di Pompei*, ed. C. Gambardella, 321–330, Napoli: La Scuola di Pitagora Editrice, 2012.)

A social survey was performed too, interviewing the tourists by a questionnaire (Figure 9.13).

The main aim of the survey was to investigate whether or not specific sound sources were expected to be heard in the area (Figure 9.14) and to discriminate the "quietness" among other conditions that can influence the judgement of the pleasantness of the area (Figure 9.15).

The social analyses confirmed the hypothesis that the sound of the technological sound sources (road traffic, aircraft flyover, and music), besides

Figure 9.13 Social survey in Pompeii.

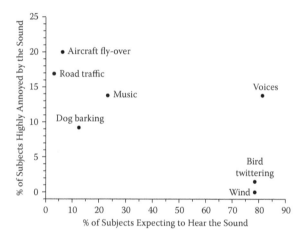

Figure 9.14 Percentage of subjects highly annoyed and auditory subjective expectation. (From Brambilla G., et al., Soundscape in the archeological area of Pompei, presented at Proceedings of 19th International Congress on Acoustics (ICA), Madrid, Spain, 2007, paper ENV-10-002.)

their sound energy level, are considered annoying and not expected. Quietness is not the most important condition for a judgement of pleasantness of the area.

Meetings with archaeologists and historians enabled us to obtain much information on the daily life of the habitants of Pompeii: where they used to stay, the main directions of the pedestrians and the transport of merchandise inside the town, where the market area was, the kind of craftsmen who operated in the town, and where they were allocated. All of this information was used as a basis for virtual maps of sound sources in the ancient

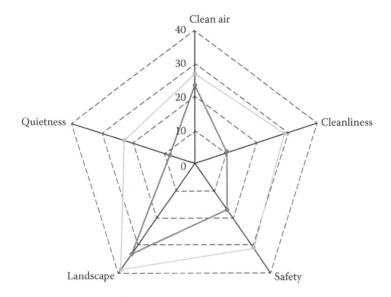

Figure 9.15 Most important aspects of the area for its pleasantness (light grey) and corresponding rating of their quality (dark grey). (From Brambilla G., et al., Soundscape in the archeological area of Pompei, presented at Proceedings of 19th International Congress on Acoustics (ICA), Madrid, Spain, 2007, paper ENV-10-002.)

Pompeii and for the tentative, for at least some sites, and still under progress, to build up virtual reality scenarios with auralization of the sound sources. The comparison between these virtual maps of the ancient times and the one reported in Figure 9.11, which refers to the actual situation, confirms the idea that large parts of the area were in the past noisy rather than quite, and this was a direct consequence of the vast expanse of the commercial town of Pompeii.

The PHS methodological analysis applied to the archaeological area of Pompeii shows that visitors do not experience fully the daily life of the ancient town, as the soundscape is absolutely not representative of the ancient situation. Technological sound sources are dominant in various parts of the area, the soundscape in other areas is strongly influenced by the behaviour of tourists, and some other areas are too quiet, referring to the past.

Efforts must be made to reconstruct a soundscape congruent with the vision of the town and part of its cultural heritage. Good results can be achieved and are under evaluation and implementation with the auralization of ancient sounds and their emission through masked loudspeakers positioned along the paths of the area (Figure 9.16), a more fluid movement and distribution of the tourists in the area obtained by attractive installations (Figure 9.17), and the control or masking of all technological sound sources outside the area.

Figure 9.16 Masked loudspeakers inside the archaeological area. (From Maffei L., et al., Valutazione del paesaggio sonoro dell'area archeologica di Pompei, in *Atlante di Pompei*, ed. C. Gambardella, 321–330, Napoli: La Scuola di Pitagora Editrice, 2012.)

9.6 FOLK FESTIVAL

Recently, the representative list of the intangible cultural heritage has been updated with the inclusion of four Italian folk festivals that belong to the network of "celebration of big shoulder-borne processional structures" (*Le Feste delle grandi macchine a spalla*) (Nardi, 2013). The folk festivals (*La Festa dei Gigli di Nola*, *Il Trasporto della Macchina di Santa Rosa a Viterbo*, *La Faradda dei Candelieri di Sassari*, and *La Varia di Palmi*) (Figure 9.18), organized in different places of central and south Italy, are expressions of ancient Mediterranean rituals that have in common a Catholic root but, more specifically, spectacular ceremonial machines that are transported on men's shoulders according to a specific rite, along paths of the historical city centres where they take place.

Although these strong common characteristics have justified a joint application to the intangible cultural heritage list, one of the festivals emphasizes intangible elements, such as music, rhythms, and sounding people movements, and these elements, together with the particular urban structure, give life to a vibrant soundscape. This festival is the *La Festa dei Gigli di Nola*, where every year in June, in the historic centre of Nola, eight obelisks (*Gigli*) are transported in procession in honor of Saint Paolino (Figure 9.19). They are wooden and papier-mâché structures with musicians located at a first stage (Figure 9.20), borne on the shoulders by a group of bearers (*paranza*) (Figure 9.21), led by a man (*Capoparanza*) who orders the movements of the *Giglio*. For a whole day, all the *Gigli* move

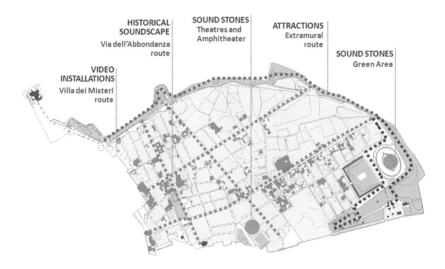

Figure 9.17 Distribution of attractive installations inside the archaeological area. (From Maffei L., et al., Valutazione del paesaggio sonoro dell'area archeologica di Pompei, in *Atlante di Pompei*, ed. C. Gambardella, 321–330, Napoli: La Scuola di Pitagora Editrice, 2012.)

Figure 9.18 The network of the "celebrations of big shoulder-borne processional structures": (a) *La Faradda dei Candelieri di Sassari*, (b) *La Festa dei Gigli di Nola*, (c) *Il Trasporto della Macchina di Santa Rosa a Viterbo*, and (d) *La Varia di Palmi*.

along a traditional path (Figure 9.22) performing spectacular manoeuvres following the rhythm of the music.

To demonstrate to what extent the perceived soundscape during the festival could be assumed as part of the cultural heritage, the PHS methodological triangulation was applied.

Audio and video recordings were carried out during the folk ceremony in order to investigate if evident soundmarks are detectable. The audio recordings tried to catch all different phases that characterize the folk ceremony,

Figure 9.19 Obelisks in the main square of Nola.

and they were performed both close to the *Gigli*, where the actors (bearers, leaders, musicians, singers) are, and along the paths, where the spectators attend the event.

The analysis in laboratory of the audio and video recordings was focused on searching specific soundmarks connected to the use of traditional instruments and musical genres, as well as to the verbal and gestural expressions that accompany the machine movements and bearers' pace.

It was found out that each of the manoeuvres of the *Gigli*, such as the *aizata*, which consists of lifting up the obelisk with a single movement performed synchronously by all the bearers under the specific order of the *Capoparanza*, or the *marcia* and *mezzo passo*, which are manoeuvres carried out by the *Gigli* to move forward along the path, or the *girata*, which is a manoeuvre consisting of a rotation of the obelisk done in very narrow streets, or several other manoeuvres, is characterized by a specific tempo and beats per minute. For instance, Figure 9.23 reports the spectrograms of the music pieces for the *marcia* with the identification of the number of beats per second (bps) with reference to the range of frequencies of percussion instruments: cymbals (a) and bass (b). The song accompanying this manoeuvre is in 6/8 tempo and with a constant execution speed of 120 bpm.

Considering the presence of eight obelisks in parade along the same path, each one with its music and crowd of rejoicing supporters, it is almost impossible to find locations not influenced by the sounds of the festival. Furthermore, the spatial conformation of the historic centre is largely made

Figure 9.20 Musicians on one of the obelisks.

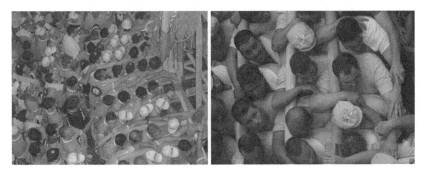

Figure 9.21 Bearers of the obelisks.

up of narrow streets and small open spaces. This situation creates a series of sound reflections and an amplification of the sound levels in the areas not involved by the passage of the *Gigli*. Indeed, the values of the sound pressure level measured during the folk ceremony range from 80.7 dB(A) in streets not affected by the *Gigli* passage to 106.5 dB(A) in positions very close to the obelisks during the manoeuvres.

Interesting and rich was the documentation that could be examined to take into account the historical criterion. The first documents of the *Festa dei Gigli* date back to 1514 and 1747 (cited in Avella, 1979, pp. 22–23). They provide information about the origin and morphological evolution of the *Gigli*. An engraving (Avella, 1979, p. 33) dated at the end of 1700 shows

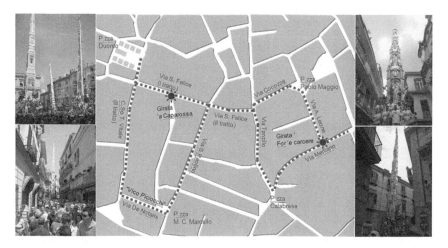

Figure 9.22 Route of the obelisks.

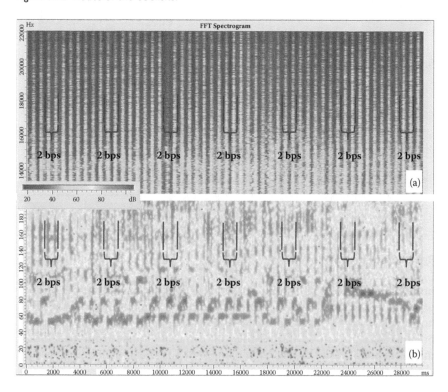

Figure 9.23 Spectrograms of the music pieces for the *marcia* with the identification of the number of beats per second (bps) in the frequency range of cymbals (a) and bass drum (b).

some pyramidal tower-shaped machines borne on men's shoulders. Seated at the first level of each machine, musicians played wind (flutes and horns) and percussion (cymbals) instruments (Figure 9.24).

In the following centuries and years, several documents (e.g., Quattromani [1838] in Esposito [1995, pp. 183–184] and Gregorovius [1853] in Avella [1979, pp. 24–26]) report descriptions of the festival, and most of them stress the role of the music and rhythms.

Other documents (Avella, 1979) provide information on the persistence and evolution of some elements found in the ancient documents.

The descriptions testify to a well-established tradition based on a language code among *Capoparanza*, *paranza*, and musicians, made of vernacular expressions, gestures, and sounds.

Up to the 1950s, the musicians were still placed above the *Gigli* and the music was played by wind and percussion instruments. New musical instruments were introduced later (e.g., sax and trombone in the 1970s or guitars and keyboards more recently), and singers have joined the musicians; however, the music is still composed or arranged in 2/4 or 6/8 times, as it is easier to associate with the movements (Figure 9.25).

The social criterion was examined by means of a survey performed during the festival on a sample of 120 subjects equally selected among bearers (actors) and spectators (locals and tourists).

The survey confirmed several preliminary assumptions valid for both actors and spectators: the sense and stimuli principally involved in the perceptual experience of the folk ceremony are the hearing, the music, and the rhythm of the *Gigli* as the most important auditory stimuli and the manoeuvres of the *Gigli* as the predominant visual stimulus.

Figure 9.24 Engraving of *Festa dei Gigli* dated the end of 1700. (From Avella L., *La Festa dei Gigli dalle origini ai nostri giorni* [*La Festa dei Gigli from the Origins to the Present Day*], Rome: Libreria Editrice Redenzione Napoletana, 1979, p. 33.)

Figure 9.25 Pictures showing the persistence of sound elements over years: (a) the 1910s (from Esposito S., *I Gigli di Nola: La Città ... la Festa [Nola's Gigli: The City ... the Ceremony]*, 14th annual, Nola: Tipo-lito "G.Scala," 1995, p. 151); (b) the 1950s (from Esposito, S., *Annali della Festa dei Gigli [Festa dei Gigli Annals]*. 2nd part. [Placeless]: Novi Italia, 1994, p. 115); (c) the 1960s (from Esposito S., *I Gigli di Nola: La Città ... la Festa [Nola's Gigli: The City ... the Ceremony]*, 14th annual, Nola: Tipo-lito "G.Scala," 1995, p. 211); (d) a *Capoparanza* leading the *Giglio* with the traditional whistle and stick in the 1970s (from Avella L., *La Festa dei Gigli dalle origini ai nostri giorni [La Festa dei Gigli from the Origins to the Present Day]*, Rome: Libreria Editrice Redenzione Napoletana, 1979, p. 61); (e) the 1970s (from Avella L., *La Festa dei Gigli dalle origini ai nostri giorni [La Festa dei Gigli from the Origins to the Present Day]*, Rome: Libreria Editrice Redenzione Napoletana, 1979, p. 61); (f) the 1990s (from Esposito S., *I Gigli di Nola: La Città ... la Festa [Nola's Gigli: The City ... the Ceremony]*, 12th annual, Nola: Tipo-lito "G.Scala," 1993, p. 138); and (g) 2011.

Furthermore, the survey showed that despite the high noise levels recorded during the event, the involvement of the population is still high, as in the past. The deafening atmosphere does not annoy people, as they are in tune with the music and the event.

The application of the PHS methodological triangulation showed that the folk festival *La Festa dei Gigli di Nola* is characterized by soundmarks that can be recognized, now, as in the past, besides some technological evolution, by the structure of the music associated with the *Gigli* manoeuvres and the specific commands given by the *Capoparanza* with vernacular expressions. The people involved as actors and spectators also recognize the strong connotation of the sound component of this folk festival and consider it, now, as in the past, predominant and attractive in the overall atmosphere. All this makes the soundscape during the event part of the intangible cultural value of the folk festival.

9.7 TOOLS FOR THE SOUNDSCAPE SAFEGUARD

Once the soundscape of a site or an event has been identified, through a PHS methodological analysis, as part of the cultural heritage of the site or event, then it is the responsibility and duty of the policy makers directly and of the community indirectly to implement all actions to safeguard and value the soundscape.

It is a responsibility because according to the operational guidelines for the implementation of the World Heritage Convention (UNESCO, 2013),"the cultural and natural heritage is among the priceless and irreplaceable assets, not only of each nation, but of humanity as a whole. The loss, through deterioration or disappearance, of any of these most prized assets constitutes an impoverishment of the heritage of all the peoples of the world."

It is a duty because (UNESCO, 2013) "all properties inscribed on the World Heritage List must have adequate long-term legislative, regulatory, institutional and/or traditional protection and management to ensure their safeguarding." This concept can be extended, however, to all sites and events whose cultural heritage is relevant besides their inclusion or not on the World Heritage List.

The protection, conservation, and valorisation of tangible and intangible cultural heritage, in such a way that it could be appreciated by present generations and passed on to future generations as their rightful inheritance, need the draft of a management plan (MP) (UNESCO, 2013; Solar, 2003), in which all strategies and actions to be implemented over time are specified. Through an MP, the authorities responsible for site management and local and national communities to which the sites belong commit themselves to an active protection, preservation, and development compatible with the cultural identities of local communities.

An MP is a strategic and operational tool that fixes flexible set of operating rules, procedures, and project ideas, whose impact is evaluated, and which involves in its implementation all stakeholders. An MP should be able to evolve and incorporate the updates and changes with the changing circumstances and the evolution of the corresponding environment.

Although an MP "depends on the type, characteristics and needs of the cultural heritage and may vary according to different cultural perspectives, the resources available and other factors" (UNESCO, 2013), the common elements of an effective management plan could include (Italian Ministry of Cultural Heritage and Activities, 2004) a knowledge project, a protection and conservation project, a strategic project for the valorisation of a local cultural and economic system, and a control and monitoring project.

The knowledge project should (1) catalogue the tangible and intangible characteristics of the cultural heritage; (2) collect information on conservation and risk status; on available legal, regulatory, and planning tools; on level of territorial infrastructure and accessibility; on tourist attraction; and on financial resources; and (3) identify the potentialities.

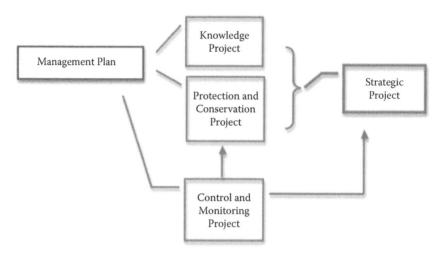

Figure 9.26 Scheme of the management plan.

The protection and conservation project should identify the safeguard measures for each single cultural heritage element to be protected and elaborate guidelines for adapting the urban planning to the needs of heritage preservation.

On the basis of the data acquired in the knowledge project and considering the limitations introduced by the protection and conservation project, the strategic project for the valorisation of a local cultural and economic system should introduce actions to emphasize the tourist potentialities, considering, however, that the variety of ongoing and proposed uses must be ecologically and culturally sustainable and must contribute to the quality of life of communities concerned. Finally, the control and monitoring project has the aim to highlight critical situations and get information on the evolution phenomenon (Figure 9.26).

The above general description of the various projects can be specified with regards to the soundscape as part of the cultural heritage.

The knowledge project could consider the constitution, at a municipal, provincial, or regional level, and in analogy with what is done for cultural heritage buildings, of a soundscape *cadastre* whose data are available and can be consulted by the local population, tourists, and stakeholders. The soundscape *cadastre* can be composed of high-quality spatial audio recordings taken in the cultural heritage sites or during the cultural heritage events. The sound recordings can be performed by means of the ambisonic (Gerzon, 1973) technique using, for example, a four capsule soundfield-type microphone that allows us to record the surrounding sound in any possible direction. By choosing an adequate loudspeaker setup or, alternatively, headphones by means of binaural techniques, after the relevant audio processing, it is possible to immerse the listener into a full acoustic field. To provide an enhanced sensorial impression, parallel video recordings can be played back, together with the sound recordings. In conjunction with audio and video information, the soundscape *cadastre* should store a description of the soundscape with information on the characteristic of sound sources and of the relative soundmarks, the data on energetic and psychoacoustic parameters, and the results of social surveys performed during the sound recordings and among the local population and tourists, with the aim of collecting their multisensorial appreciation of the site or event. The soundscape *cadastre*, which not only is a knowledge instrument, but also could represent a cultural value for the future, should be updated periodically, allowing us to keep track of the soundscape evolution during the time.

The protection and conservation project should verify and avoid the negative effect of modification of the original soundscape due to the introduction of new urban or morphological elements or new sound sources or new uses of the space. This can be achieved by the auralization techniques (Vorländer, 2011) that enable us to build up new virtual audio scenarios. These, before any real modification, can be compared to the original soundscape, and the judgement of a jury of experts, and not experts on the acceptability of the modification, with reference to the time evolution of uses, is also acquired. The test can be enriched and rendered more ecologically valid (Guastavino et al., 2005) with use of immersive virtual reality techniques, in which the jury can live and examine the new scenario in a multisensorial way.

The same tests can be used to elaborate the strategic project for the valorisation of a local cultural and economic system. In this case, the projects of new audio scenarios do not suffer, but they are specifically elaborated by stakeholders and then tested. In this case, juries can also be composed by groups of tourists.

Finally, the control and monitoring project should introduce systems that can acquire in real time information on the soundscape and compare this with information conserved in the soundscape *cadastre*. This information constitutes acoustic descriptors acquired with fixed monitoring systems placed in strategic positions of the sites and people's perceptual reactions, acquired through mobile systems whose use is nowadays more common among the local population and tourists.

REFERENCES

Avella L. (1979). *La Festa dei Gigli dalle origini ai nostri giorni* [*La Festa dei Gigli from the Origins to the Present Day*]. Rome: Libreria Editrice Redenzione Napoletana.

Basturk S., Carafa R., Maffei L. (2010). The validation of architectural and acoustic projects to transform ecclesiastical architecture in auditoria for Concert music. Presented at Proceedings of Euroregio 2010, Congress on Sound and Vibration, Ljubljana, Slovenia.

Basturk S., Maffei L., Masullo M. (2012). Soundscape approach for a holistic urban design. Presented at Proceedings of AESOP—26th Annual Congress, Ankara, Turkey.

Basturk S., Maffei L., Perea Pérez F., Ranea Palma Á. (2011a). Multisensory evaluation to support urban decision making. Presented at Proceedings of International Seminar on Virtual Acoustics—ISVA 2011, Valencia, Spain.

Basturk S., Ranea Palma Á., Perea Pérez F., Maffei L. (2011b). Análisis multisensorial de la contaminación acústica en espacios urbanos. Presented at Proceedings of TecniAcustica 2011, Cáceres, Spain.

Botteldooren D., Andringa T., Aspuru I., Brown L., Dubois D., Guastavino C., Lavandier C., Nilsson M., Preis A. (2013). Soundscape of European cities and landscape: Understanding and exchanging. In *Soundscape of European Cities and Landscapes*, ed. J. Kang, K. Chourmouziadou, K. Sakantamis, B. Wang, Y. Hao, 36–41. Oxford: Soundscape-COST.

Brambilla G., De Gregorio L., Maffei L., Masullo M. (2007a). Soundscape in the archeological area of Pompei. Presented at Proceedings of 19th International Congress on Acoustics (ICA), Madrid, Spain, paper ENV-10-002.

Brambilla G., De Gregorio L., Maffei L., Yuksel Can Z., Ozcevik A. (2007b). Comparison of the soundscape in the historical centres of Istanbul and Naples. Presented at Proceedings of the 36th International Congress on Noise Control Engineering, Internoise 2007, Istanbul, Turkey, paper IN07-334.

Brambilla G., Maffei L. (2010). Perspective of the soundscape approach as a tool for urban space design. *Noise Control Eng. J.*, 58(5), 532–539.

Brambilla G., Maffei L. (2011). Soundscape heritage: An evolving value to preserve and archive? Presented at Think Tank: Soundscape as a Part of Cultural Heritage [organized and supported by COST Action TD0804], Capri, Italy.

Brambilla G., Maffei L., Di Gabriele M., Gallo V. (2013). Merging physical parameters and laboratory subjective ratings for the soundscape assessment of urban squares. *J. Acoust. Soc. Am.*, 134(1), 782–790.

Drettakis G., Roussou M., Reche A., Tsingos N. (2007). Design and evaluation of a real-world virtual environment for architecture and urban planning. *Presence Teleoperators Virtual Environ.*, 16(3), 318–332.

Di Gabriele M., Maffei L., Aletta F. (2010). Urban noise mapping based on emotional dimensions. Presented at Proceedings of Euroregio 2010, Congress on Sound and Vibration, Ljubljana, Slovenia, paper 075.

Esposito S. (1993). *I Gigli di Nola: La Città la Festa* [*Nola's Gigli: The City ... the Ceremony*]. 12th annual. Nola: Tipo-lito "G.Scala."

Esposito S. (1995). *I Gigli di Nola: La Città la Festa* [*Nola's Gigli: The City ... the Ceremony*]. 14th annual. Nola: Tipo-lito "G.Scala."

Esposito S. (1994). Annali della Festa dei Gigli (Festa dei Gigli Annals) (in Italian). 2nd part. [Placeless]: Novi Italia.

Funkhouser T., Min P., Carlbom I. (1999). Real-time acoustic modeling for distributed virtual environments. Presented at Proceedings of 26th Annual Conference on Computer Graphics and Interactive Techniques—SIGGRAPH 1999, Los Angeles.

Gerzon M. A. (1973). Periphony: With height sound reproduction. *J. Audio Eng. Soc.*, 21(1), 2–10.

Gregorovius F. (1853). *Passeggiate in Campania e in Puglia* [*Excursions in Campania and in Puglia*], 59–70. 2nd ed. Roma: Edizione Spinosi. Cited in Avella L. (1979). *La Festa dei Gigli dalle origini ai nostri giorni* [*La Festa dei Gigli from the Origins to the Present Day*], 24–26. Roma: Libreria Editrice Redenzione Napoletana.

Guastavino C., Katz B. F. G., Polack J. D., Levitin D. J., Dubois D. (2005). Ecological validity of soundscape reproduction. *Acta. Acust. united Acust.*, 91, 333–341.

Guillén J. D., López Barrio I. (2007). Importance of personal, attitudinal and contextual variables in the assessment of pleasantness of the urban sound environment. Presented at Proceedings of 19th International Congress on Acoustics (ICA) 2007, Madrid, paper ENV-10-009.

Haverkamp M. (2007). Essentials for description of cross-sensory interaction during perception of a complex environment. Presented at Proceedings of the 36th International Congress on Noise Control Engineering, Internoise 2007, Istanbul, Turkey.

Hussein A. (2009). The use of triangulation in social sciences research: Can qualitative and quantitative methods be combined? *J. Comparative Social Work*, 1, 1–12.

Italian Ministry of Cultural Heritage and Activities. (2004). Model for the management plan of properties included in the World Heritage List. Paestum, May 24–25.

Lercher P., Schulte-Fortkamp B. (2013). Soundscape of European cities and landscapes: Harmonising. In *Soundscape of European Cities and Landscapes*, ed. J. Kang, K. Chourmouziadou, K. Sakantamis, B. Wang, Y. Hao, 120–127. Oxford: Soundscape-COST.

Maffei L. (2008). Urban and quiet areas soundscape preservation. Presented at Proceedings of VI Congreso Iberoamericano de Acústica—FIA 2008, Buenos Aires, FIA2008-003.

Maffei L. (2010). Soundscape approach in urban renewal: Parks and areas of acoustic quality. In *Proceedings of DAGA 2010, 36th Jahrestagung fur Akustik*, Berlin, pp. 51–52.

Maffei L., Di Gabriele M., Brambilla G., De Gregorio L., Natale R. (2012). Valutazione del paesaggio sonoro dell'area archeologica di Pompei. In *Atlante di Pompei*, ed. C. Gambardella, 321–330. Napoli: La Scuola di Pitagora Editrice.

McDonald P., Rice H., Dobbyn S. (2009). The sound of the City: Auralisation of community noise data. Presented at Proceedings of Internoise 2009, Ottawa.

Nardi P. (2013). Dialogare in rete—Le feste delle grandi Macchine a spalla italiane [Dialoguing on the net: The celebrations of the Italian big shoulder-borne processional structures], *SITI*, 12, 118–122.

Quattromani G. (1838/1995). La Festa di San Paolino a Nola [Saint Paolino ceremony in Nola]. In *I Gigli di Nola: La Città la Festa* [*Nola's Gigli: The City ... the Ceremony*], ed. S. Esposito, 183–184. 14th annual. Nola: Tipo-lito "G.Scala."

Raichel D. R., Brooks B. M., Lubman D. (2004). Archaeological acoustics: A guide to trends in community noise levels. *J. Acoust. Soc. Am.*, 115(5), 2622.

Schafer R. M. (1977). *The Tuning of the World*. Toronto: McClelland & Stewart.

Solar G. (2003). Site management plans: What are they all about? *World Heritage*, 31, 22–23.

Tsingos N., Gallo E., Drettakis G. (2004). Perceptual audio rendering of complex virtual environments. Presented at Proceedings of 31st Annual Conference on Computer Graphics and Interactive Techniques—SIGGRAPH 2004, Los Angeles.

UNESCO. (1954). Convention for the Protection of Cultural Property in the Event of Armed Conflict with Regulations for the Execution of the Convention, The Hague.

UNESCO. (1972). Convention Concerning the Protection of the World Cultural and Natural Heritage, Paris.

UNESCO. (1995). Report of the World Heritage Committee (19th session), Berlin (WHC-95/CONF 203/16).

UNESCO. (1997). Report of the World Heritage Committee (21st session), Naples, Italy (WHC-97/CONF 208/17).

UNESCO. (2001). Convention on the Protection of the Underwater Cultural Heritage, Paris.

UNESCO. (2003). Convention for the Safeguarding of Intangible Cultural Heritage, Paris.

UNESCO. (2013). Operational guidelines for the implementation of the World Heritage Convention (WHC). July.

Vorländer, M. (2008). Virtual acoustics: Opportunities and limits of spatial sound reproduction for audiology. Presented at Proceedings of DGMP Tagung 2008, Oldenburg, Germany.

Vorländer M. (2011). *Auralization: Fundamentals of Acoustics, Modelling, Simulation, Algorithms and Acoustic Virtual Reality*. Berlin: Springer.

Yeasmin S., Ferdousour Rahman K. (2012). "Triangulation" research method as the tool of social science research. *BUP J.*, 1(1), 154–163.

Chapter 10

Applied Soundscape Practices

*Lisa Lavia,[1] Max Dixon,[2] Harry J. Witchel,[3]
and Mike Goldsmith[4]*

[1]Noise Abatement Society, United Kingdom

[2]Independent Consultant, United Kingdom

[3]Brighton and Sussex Medical School, United Kingdom

[4]Acoustic Consultant, United Kingdom

CONTENTS

10.1 INTRODUCTION

Soundscape studies cover a very broad and rapidly evolving field. As we have learned repeatedly throughout this book, the recent management of the acoustic environment has predominantly been concerned with reducing or masking sound levels. While science will continue to investigate and find better ways to measure, hone, and develop soundscape management, certain key determinants exist now upon which we can establish a holistic foundation for the future development of the discipline: sound quality measurement, psychobiological factors, and sociological indicators.

While there are still relatively few concrete soundscape improvement projects triangulating data from all of these factors, the examples in this chapter represent a range of practical soundscape applications recently engaged in by practitioners associated with the European Union Cooperation in Science and Technology (COST) Action TD0804 (http://soundscape-cost.org/), having a focus on landscape, acoustic ecology, and the built environment (Lavia and Dixon, 2013).

While not a comprehensive survey of the state of the art, each example chosen illustrates how the approach offers the opportunity for interdisciplinary and multidisciplinary work, bringing together science, medicine, social studies, and the arts—combined, crucially, with analysis, advice, and feedback from the users of the space as the primary experts in any environment—to find creative and responsive solutions for the acoustic design and protection of places (Figure 10.1).

In each of the following examples, the relevant soundscape intervention/ management types are described: introducing sounds to the soundscape, utilization of sounds that already exist in the location, assessment methodology, incorporation of sonic art installations, utilization of noise control elements, and design alterations. The types are aligned with the elements of the soundscape protocol (Figure 10.1) to identify how each study applied the concepts.

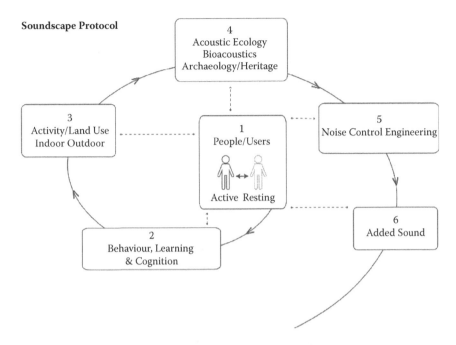

Soundscape Protocol

Figure 10.1 Elements of the soundscape protocol. (Copyright © Noise Abatement Society, UK.)

10.2 PURPOSE-DESIGNED SOUNDSCAPE PROJECTS

10.2.1 Sounding Brighton: A Practical Approach to Soundscape-Based City Planning

10.2.1.1 Synopsis

Project name: Sounding Brighton
Location: City of Brighton and Hove, United Kingdom

10.2.1.2 Researchers' Names and Affiliations

Lisa Lavia: lisa.lavia@noise-abatement.org
Matt Easteal: Matt.Easteal@brighton-hove.gov.uk
Simon Bannister: Simon.Bannister@brighton-hove.gov.uk
Harry J. Witchel: h.witchel@bsms.ac.uk
Jian Kang: j.kang@sheffield.ac.uk
Östen Axelssön: oan@psychology.su.se
Max Dixon: maxdixon1@hotmail.co.uk
Donna Close: D.Close2@brighton.ac.uk
Francesco Aletta: f.aletta@sheffield.ac.uk
Carina Westling: c.e.i.westling@sussex.ac.uk

10.2.1.3 Soundscape Intervention/Management Types

Introduction of sounds to the soundscape: Sounding Brighton is a series of projects including trial introductions of musical and other recorded sounds into the acoustic environment. First, a diverse range of sounds was tested in a three-dimensional (3D) outdoor soundscape installation to assess the potential for reducing antisocial behaviours among nightclub goers. Second, reactions of foot tunnel users to the introduction of a range of types of recorded music were assessed.

Incorporation of sonic art installations: The first project was carried out in the context of White Night, an all-night, citywide arts and cultural festival that included sonic art installations in other locations. The second project could have been regarded as a sonic art installation itself.

Utilization of sounds that already exist in the location: In the first project, artificial sound input was continuously edited manually in response to changes in the existing ambient environment. In the second project, different levels of musical sound were selected, having regard for typical conditions in a foot tunnel, but were not ambient responsive.

Assessment methodology: In the first (pilot) project, assessment was both objective and subjective, but also included engagement with citizen audiences via live AV feeds in a nearby public hall. In the second project, citizen reactions were assessed largely by analysis of video recordings. The third project in the series was a social survey including citizen's views on the appropriateness of sounds in their favourite location. The fourth project is piloting and seeking to embed urban sound planning principles in the development of the city.

10.2.1.4 How the Project Has Contributed to Development of Soundscape Concepts

The pilot studies comprising Sounding Brighton represent the most complete collection of soundscape-led studies involving a single city administration. The pilots each utilize a systematic approach to soundscape interventions and design in a specific local context. Those involved in the pilots have acknowledged that limited resources were available for further evidence gathering and analysis, and have noted that additional resources to build upon the existing findings would be a priority in future developments of the project.

10.2.1.5 Project Overview and Analysis

In 2010 the City of Brighton and Hove in England invited the UK Noise Abatement Society (NAS) to propose solutions for its ongoing issues with night noise disturbance from the late-night economy, which was adversely affecting local residents. The result was the Sounding Brighton project series (Lavia and Bennett, 2011), which was supported by the European Union COST Action TD0804 network and members of the International Organization for Standardization (ISO 12913-1:2014) on soundscape (ISO 12913-1), in addition to Brighton and Hove City Council and the UK Noise Abatement Society.

The project's overall aim was to modify the soundscapes of affected areas, and it incorporated international workshops to collate best practice in the field and use it to develop projects in Brighton and Hove, both to gather evidence and test and develop practical solutions (http://soundscape-cost.org/).

10.2.1.6 2011: Night Noise Intervention Pilot Study (Lavia et al., 2012b)

A particular focus was White Night, an all-night arts and cultural festival, and an experimental night noise intervention exercise named West Street Story (Lavia et al., 2012b) that was co-commissioned by White Night NAS and Brighton and Hove City Council based on consultation with local residents, the council and club owners, and a community survey. The premise for the experiment was based on research compiled by Dr. Harry Witchel in his book *You Are What You Hear: How Music and Territory Make Us Who We Are* (Witchel, 2010), which proposes that music can engender territorial and pleasure responses, depending on whether listeners like what they hear. Artificial soundscapes can make places enjoyable and welcoming, and even help create a place of togetherness.

To this end, a soundscape was created by Martyn Ware of the Illustrious Company (http://illustriouscompany.co.uk/performance/west-street-story). The installation was situated in part of West Street (Figure 10.2), in the heart of Brighton's night life, and designated the most dangerous street in the city. It used two rows of speakers to create a soundscape through which people walked (http://soundscape-cost.org/index.php?option=com_content&view=article&id=58:west-street-story&catid=35:soundscape-practices&Itemid=8) (Lavia et al., 2012c). This soundscape presented a contrast to the raucous disharmony so frequently heard in lively areas at night, for the benefit of visitors to the area and residents, as well as those exiting the clubs.

Audio from this project, together with videos of crowds in the area, was discussed at the Come Together event in Brighton University's Sallis Benney Theatre, facilitated by Driftwood Productions (Lavia and Bennett, 2011). There, psychobiologist and communications expert Dr. Harry Witchel, from Brighton and Sussex Medical School, managed three interactive master classes about body language, music, and social territory. These enabled participants to see the effects of acoustic environments on the body language and behaviour of people in general, as well as those filmed during White Night.

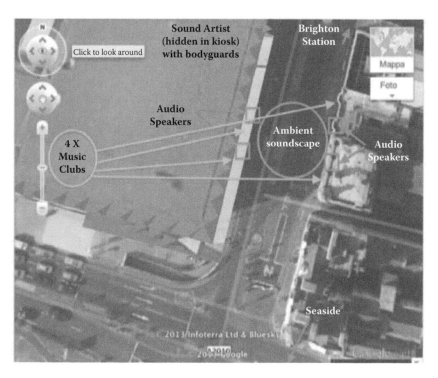

Figure 10.2 Three-dimensional ambient soundscape night noise intervention pilot study staged during an all-night arts and cultural festival, White Night, in Brighton, UK, on October 29, 2011. (From Lavia, L., et al., *Sounding Brighton: Practical approaches towards better soundscapes*, presented at Internoise 2012, New York, 2012. Copyright © Google.)

Other related events, co-curated by the White Night NAS and Brighton and Hove City Council (Lavia and Bennett, 2011) included Creation Power, in which Dr. John Drever from Goldsmiths College, University of London, illustrated how designers of gadgets and machines, and the individuals using them, should be aware of the impact of sounds associated with these products and the ways in which they affect people and spaces (Drever, 2012). Another event was Brighton Remixed, featuring a soundscape installation based on listeners' sonic experiences of Brighton and Hove. It was designed by the NAS and Esther Springett (http://www.estherspringett.com/teaching_facilitating/brighton-remixed/), a sound artist and facilitator, working with Dv8 Training Brighton, who provide media-based training for young people. Esther helped a group of 16- to 18-year-olds to explore their own soundscapes, listen in new ways, and learn technical skills to help them access opportunities in the creative industries.

Trial results were encouraging. In particular, in the West Street Story experiment (Lavia et al., 2012b), people laughed, hugged, and danced spontaneously—a marked departure from the normal tensions, aggressions, and frequently intimidating atmosphere the area is known for. Providing feedback on the event, police commented on how much quieter the West Street area was, so much so that they were confident enough to redeploy resources elsewhere in the city; 96% of the audience surveyed felt safe in the city that night, compared to 50% on a usual night. The Alcohol Programme Board in the city are interested in supporting further work building on the lessons from West Street Story, as the outcomes of the project have come to be known (Brighton and Hove City Council).

10.2.1.7 2013: Place Making and Antisocial Behaviour Intervention (Witchel et al., 2014)

The results of the White Night project suggest soundscape management can be a valuable resource in helping mitigate antisocial behaviour; however, more detailed investigations into soundscape management in real-world environments are required to validate this approach. A more scientifically controlled investigation of the potential of artificial soundscapes to improve public safety was therefore undertaken. The investigation, called West Street Tunnel (Witchel et al., 2013) and designed by the NAS and Dr. Harry Witchel, was based on objective indicators, including walking speed and loitering. Such indicators could potentially be combined with others to infer more general prevailing psychological states, in particular an enhanced sense of security.

10.2.1.7.1 Loitering

Gauged from anecdotal reports, general loitering activities seem to be one of the most prominent behaviours leading to feelings of unease and concerns over security in the tunnel. We have previously characterized loitering in territorial terms, as a phenomenon whereby individuals will remain in the tunnel for prolonged periods of time, displaying behaviours that advance ownership claims over its attractive territorial affordances. These territorial behaviours are often perceived as exclusive or defensive in nature, discouraging other members of the public from entering what they perceive as someone else's territory, or causing feelings of discomfort while passing through such a region. As one of the prevailing aims motivating our investigations has been to determine possible public safety interventions, we performed some preliminary tests to help determine the scope for possible soundscape interventions on this front, and whether they might reduce loitering.

In order to measure loitering, we began with the video data captured over two consecutive Saturdays when the music intervention was active (between the hours 0000 and 0700 and 1900 and 2400). Our experimental setup, including motion-activated video recording, enabled us to selectively record the entire set of people travelling through the tunnel during this time period (a set of $n = 559$ films, each having a minimum of one person moving in the tunnel). From this initial set, we looked exclusively at films of file size > 6 megabytes ($n = 114$), thus filtering out those films showing activity in the tunnel for less than 60 s—for people walking naturally through the tunnel, they would be on camera for ~20 s, resulting in a continuous film of size 1–3 megabytes. Often these larger video files would merely display multiple groups or individuals travelling consecutively through the tunnel (rather than one individual being in the tunnel for an extended period), thus activating the video recorder for longer, while no loitering behaviours were in fact occurring. Therefore, we subjected these films to manual analysis, and tracked people across multiple films where necessary, in order to gauge loitering levels, defining loitering in this instance as lingering in the tunnel for longer than 60 s.

Using this setup, we found little loitering activity, with only 16 instances of either individual or group loitering being detected across the two 12 h periods. Loitering would most commonly occur in groups (of size 2–11), with an average group size of 3 people. This gave us a total of 49 individuals detected loitering, much less than we had anticipated for a regular weekend period. We found a median loitering time of 4 min 35 s, and the range of activities engaged in during these loitering periods is shown in Figure 10.3. From these measurements we can thus give a minimum estimate of 8 loitering instances, or 25 individual loiterers, per day. The observed loitering instances are shown on Figure 10.3, including the time spent in the tunnel as well as the dominant activities performed while loitering; as expected, the long periods (>20 min) spent loitering were associated with sitting and lying down, with really long periods in the tunnel spent sleeping.

As a surrogate measure of whether music might deter or encourage loitering, we looked at whether loitering instances began or ended on a particular piece of music, based on the assumption (put forward in Witchel [2010]) that if people's motivations for remaining in the tunnel are on the borderline between wanting to stay or go, the influence of a piece of music that they find counterterritorial will be enough to initiate a movement to leave during that piece of music. The manual analysis of films enabled us to keep track of the music playing in the tunnel at the start and end points of loitering events (as indicated on Figure 10.4). Here, we found that classical music (Handel) and jazz were associated with the beginning (entry) of fewer loitering episodes than would be expected by random chance, and likewise, classical music was associated with more exits than expected if these events had occurred randomly with respect to the music. These data suggest that, compared with the other music, we might suspect that classical music functions as a loitering deterrent, while silence (overrepresented during entry) might be comparatively welcoming for loiterers. Plainly, 16 instances are a very small set of loitering episodes to draw inferences from, but there may be enough data within the entire corpus to draw firmer conclusions.

Group Number	Number of Loiterers	Minutes Loitering	Seconds Loitering	Main Activities	Music at Start	Music at End
1	3	2	8	Talking	Jazz	Ambient/Dance
2	2	2	14	Talking and hugging	Jazz	Silence
3	5	4	34	Talkin with friends	Ambient/Dance	Jazz
4	11	1	30	Playing group games	Silence	Ambient/Dance
5	2	1	24	Dancing	Ambient/Dance	Ambient/Dance
6	2	4	35	Sitting and smoking	Ambient/Dance	Classical
7	2	50	0	Sleeping	Jazz	Classical
8	4	6	44	Dancing and sitting	Silence	Ambient/Dance
9	6	1	29	Talking, look at posters, drinking	Ambient/Dance	Ambient/Dance
10	2	24	25	Seating, eating and smoking	Ambient/Dance	Classical
11	1	207	39	Sleeping	Ambient/Dance	Silence
12	1	1	30	Smoking	Ambient/Dance	Classical
13	2	11	56	Talking	Silence	Silence
14	2	35	20	Sitting and talking	Silence	Classical
15	2	15	37	Talking	Ambient/Dance	Jazz
16	2	1	57	Talking	Classical	Jazz

Figure 10.3 Raw loitering data for two Saturdays in West Street Tunnel during a night noise intervention experiment. (From Witchel, H.J., et al., Music interventions in the West Street Tunnel in Brighton: A community safety and night-noise soundscape intervention pilot: Preliminary report, Brighton and Hove, UK: Brighton and Hove City Council, 2014.)

	% Playlist	Enter During	Expected	Exit During	Expected
Classical	18.7%	1	3.00	5	3.00
Jazz	28.1%	3	4.50	3	4.50
Ambient/Dance	39.1%	8	6.25	5	6.25
Silence	14.1%	4	2.25	3	2.25
TOTAL	100.0%	16	16.00	16	16.00

Figure 10.4 Comparison of actual loitering events to expected loitering events if the events were randomly distributed without reference to music. Because the music excerpts were not of the same length, in order to determine whether music encouraged exiting from the tunnel or discouraged entering (and remaining) in the tunnel, we calculated the expected number of entry and exit events based on the total number of loitering events multiplied by the percentage of time in the playlist occupied by that song/excerpt. If music had no effect on loitering, then the distribution of how many loitering events occurred would mirror the percentage of the playlist spent on each excerpt (i.e., the numbers in the expected column would be identical to the entry and exit columns). If the number of entries is lower than expected (e.g., classical and jazz), then commencement of loitering events was potentially discouraged during this music, and if music during entry (and deciding to loiter) was higher than expected (silence), then loitering might be comparatively encouraged (i.e., this music was welcoming for loiterers). Likewise, if the number of exits is higher than expected, then this might indicate that the music discouraged loitering (e.g., classical). (From Witchel, H.J., et al., Music interventions in the West Street Tunnel in Brighton: A community safety and night-noise soundscape intervention pilot: Preliminary report, Brighton and Hove, UK: Brighton and Hove City Council, 2014.)

Our results indicate overall a smaller number of loitering instances than we might have expected. They provide some useful preliminary indications of the levels of loitering we might anticipate in future observations, and give us the scope for further analysis of the effect of specific music on loitering.

10.2.1.7.2 Walking Speed

We used a small playlist so that we could compare different people walking through the tunnel on the same day (even in the same hour). This allowed us to make preliminary measurements of how people responded to music versus silence, and to compare the same piece of music digitally altered to be 10% faster (but digitally corrected to maintain pitch) (Lopez-Mendez et al., 2014).

When contemporary instrumental electronica/dance music (Cirrus) was deployed, people's walking velocity through the tunnel was slower (mean = 1.23 m/s, $n = 105$) than in the absence of music (Figure 10.5; mean = 1.44 m/s, $n = 22$, unpaired t-test: $p < .0001$). Reference values from the scientific literature suggest healthy men in their 20s are thought to walk

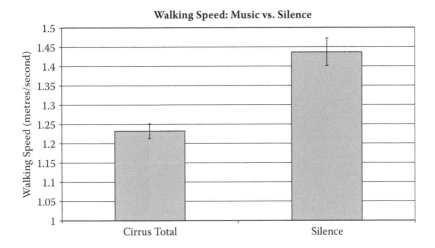

Figure 10.5 Difference in walking speed during music and silence in the West Street Tunnel experiment. (From Witchel, H.J., et al., Music interventions in the West Street Tunnel in Brighton: A community safety and night-noise soundscape intervention pilot: Preliminary report, Brighton and Hove, UK: Brighton and Hove City Council, 2014.)

comfortably at 1.393 m/s, while women in their 20s walk at 1.407 m/s (Bohannon, 1997). This suggests that, based on our measurement techniques on the multiaged tunnel cohort, during silence the passersby walked at almost exactly the scientific reference rate; however, when listening to Cirrus, people walked significantly slower than normal.

Second, the same piece of music at two contrasting tempos (all other variables held constant) resulted in significantly different walking velocities. Faster tempo music consistently made people travel quicker through the tunnel (Figure 10.6) (mean velocity = 1.27 m/s, n = 52) than did slower tempo music (mean velocity = 1.19 m/s, n = 53, unpaired t-test; $p < .05$).

The investigation was based in a pedestrian subway (Figures 10.7 and 10.8), which allowed adjustments to the environment (specifically the addition of music and colourful posters) to be easily managed, which also (as with West Street) had a history of antisocial behaviour. It was designed as a safe, automobile-free, pedestrian thoroughfare to and from the West Street district to the beach; however, the antisocial activities with which it was plagued led the council to close it at night. One goal of the work was to help make the foot tunnel a more appealing and safer environment, and thus to make way for its possible reopening during nighttime hours.

In contrast to the West Street Story (Lavia et al., 2012b) experiment, the enclosed nature of the environment and the fact that most pedestrians spent about the same amount of time meant that the dose of music per person was fairly constant.

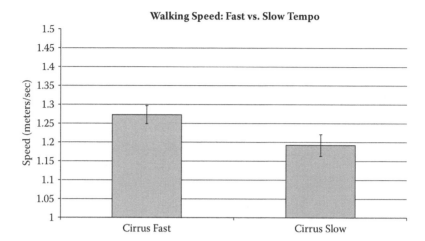

Figure 10.6 Difference in walking speed during a fast- and slow-tempo version of the same musical excerpt during the West Street Tunnel experiment. (From Witchel, H.J., et al., Music interventions in the West Street Tunnel in Brighton: A community safety and night-noise soundscape intervention pilot: Preliminary report, Brighton and Hove, UK: Brighton and Hove City Council, 2014.)

Along the floor we installed white stripes made of durable duct tape; these were positioned every 210 cm to allow for the calculation of the speeds of people moving along the tunnel (Figure 10.9).

The aim of the musical intervention was to induce the feeling that the tunnel was an inclusive environment, promoting social cohesion, in which more pro-social behaviours were considered appropriate (Witchel et al., 2014). A variety of styles and tempos of music were provided, and the effects of these and the posters on objective behavioural indicators were video recorded. Acoustic measurements were also made.

Analysis of the video data is a mixture of manual analysis and semiautomated video analysis based on computer vision (Figure 10.10). Classical music was found to diminish the surrogate measures of loitering compared to silence or other music. Faster-tempo music led to faster walking speeds than slower music. The presence of music resulted in slower walking than silence, and sometimes led to dancing in the tunnel. Given these observations, we tentatively conclude that this type of night noise intervention is technically feasible in an exposed space for months at a time, and that different kinds of music can have an immediate pro-social effect.

10.2.1.7.3 Future Work

Audio and sound recordings are being analyzed in a subsequent project with Brighton and Hove City Council, the NAS, Professor Jian Kang,

Figure 10.7 Aerial view of the tunnel's entrances. The roadway intersection is the busy King's Road (going left to right) meeting West Street (going upward). North is upward. The upper arrow points to the descending entrance to the tunnel (descending from right to left into the shadow). The lower arrow points to where the tunnel mouth opens onto the beach promenade. (From Witchel, H.J., et al., Music interventions in the West Street Tunnel in Brighton: A community safety and night-noise soundscape intervention pilot: Preliminary report, Brighton and Hove, UK: Brighton and Hove City Council, 2014. Copyright © Google.)

Dr. Francesco Aletta, and Dr. Harry Witchel (Easteal et al., 2014) in order to provide further insights as to which sound quality characteristics were aligned with which responses of the user(s) in context. This evidence will provide the opportunity to model a soundscape management approach for the specific scenario and the existing and altered environment relative to the user(s) of the space.

In addition to night noise intervention pilots, Sounding Brighton has also included other innovative applied approaches to soundscape management.

10.2.1.8 2011: Soundscape Mapping to Inform the City's Response to Localized Noise Issues (Lavia et al., 2012a)

In this experiment, the NAS conducted a soundscape-based social survey, in cooperation with Brighton and Hove City Council, local action teams (residents' groups), Dr. Östen Axelsson, and Max Dixon, in which members

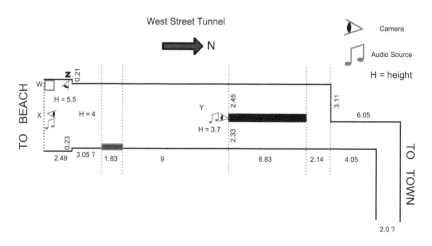

Figure 10.8 Schematic diagram of the West Street Tunnel. North is directly rightward. The tunnel slopes downward to the left (i.e., toward the beach) in the diagram. The tunnel divides into two tunnels in its middle, and the westward tunnel (upper in the diagram) has lockable metal gates at both ends. Our team installed several electronic units: (W) A locked cabinet with 240 VAC mains power. The cabinet housed the digital video recorder, the power amp, and the iPhone music presentation system. Power to the cameras also originated from the cabinet. (X) Camera 1 (north facing) + speaker in a vandalism-resistant metal cage. Located above archway (but not at its peak). (Y) Camera 3 (south facing) + speaker in a vandalism-resistant metal cage. Note that camera 3 failed shortly after installation, so no data were recorded by it. (Z) Camera 2 (southeast facing) + microphone to record ambient noise. (From Witchel et al., Preliminary report. Brighton and Hove City Council, Brighton and Hove, UK, 2014.)

of the public were invited to participate anonymously (Lavia et al., 2012a). The survey collected information of four types: (1) noise annoyance, (2) favourite outdoor location in Brighton and Hove City and the social and recreational activities that take place there, (3) what sounds are appropriate to the favourite location, and (4) demographic data. Using hierarchical cluster analysis, five categories of favourite locations, as well as five categories of recreational soundscapes, were identified.

The work raises the possibility of integrating a soundscape approach with a recent method for spatial planning and land use management, developed at the Stockholm Urban Planning Administration, through user surveys called sociotope mapping: defined as "the commonly perceived direct use values of a place by a specific culture or group" (Ståhle, 2006). In Brighton and Hove the sociotope approach is being adapted through the Sounding Brighton project to encompass soundscape, by incorporating the sounds people find appropriate to the sociotope (Table 10.1) into soundscape maps.

With this pilot study we established that the present method is useful for identifying distinct sociotopes and soundscapes, and have shown that there

Figure 10.9 Inside West Street Tunnel, looking northward. The horizontal stripes on the floor were white duct tape that required only minor maintenance every fortnight, despite being walked on by thousands of people and often being soaked in the puddles of water that accumulated in the tunnel during rainy weather. Two large poster photographs of Brighton can be seen on the walls (left panel). Note the incline upward in the north half of the tunnel; rainwater would flow down this incline and accumulate as puddles on the west side of the tunnel (right panel). (From Witchel, H.J., et al., Music interventions in the West Street Tunnel in Brighton: A community safety and night-noise soundscape intervention pilot: Preliminary report, Brighton and Hove, UK: Brighton and Hove City Council, 2014.)

Figure 10.10 Example of motion tracking of a person walking in the tunnel from Jean-Marc Odobez's lab. This tracking algorithm was used for testing entrainment to rhythm (i.e., up–down vs. side-to-side motion). (From Lopez-Mendez, A., et al., Automated bobbing and phase analysis to measure walking entrainment to music, presented at IEEE International Conference on Image Processing, Paris, 2014.)

is a statistically significant relationship between them: people agree to a large extent on the degree to which a specified set of sound sources is appropriate in a place that is suitable for a specified set of social and recreational activities.

With regards to urban planning, the method utilized for this study could be used to identify and decide acoustic objectives for an area depending on what social and recreational activities it is intended for, so

Table 10.1 Contingency Table for Five Sociotopes and Five Soundscapes

Sociotope	Soundscape					
	Urban Nature	Distant Nature	Urban	Seaside	Urban Beach	Totals
Beach and seaside	41 (49.5)	15 (32.7)	31 (31.9)	44 (20.6)	12 (8.3)	143
City park	46 (40.2)	34 (26.5)	27 (25.9)	4 (16.7)	5 (6.7)	116
Peri-urban recreation area	34 (19.1)	20 (12.6)	1 (12.3)	0 (7.9)	0 (3.2)	55
My space	7 (13.5)	17 (8.9)	12 (8.7)	2 (5.6)	1 (2.3)	39
Downtown city	4 (9.7)	1 (6.4)	14 (6.2)	5 (4)	4 (1.6)	28
Totals	132	87	85	55	22	381

Source: Lavia, L., et al., Sounding Brighton: Practical approaches towards better soundscapes, presented at Internoise 2012, New York, 2012.

Note: Numbers in parentheses are the values expected by chance. Results from the soundscape survey of residents in Brighton and Hove based on Swedish sociotope mapping. They show that the beach and seaside sociotope was strongly associated with the seaside and urban beach soundscapes and weakly associated with the distant nature soundscape; the city park sociotope was associated with the urban nature and distant nature soundscapes, but very weakly with the seaside soundscape, and so forth for each category.

that soundscape can be taken into account early on in the planning stage (Axelsson, 2011). A next step could be to develop soundscape mapping to complement sociotope maps for use by city administrations, urban planners, and architects.

10.2.1.9 2012: Lobbying Local, National, and International Stakeholders

A public exhibition initially presented within Brighton Town Hall, UK (2012) and subsequently at Vitoria-Gasteiz, Spain (2012); the AIA-DAGA Conference on Acoustics, Merano, Italy (2013); Naples, Italy (2013); Internoise Innsbruck, Austria (2013), showcased the latest in innovative soundscape practice and solutions (Lavia and Dixon, 2013). It featured applied soundscape projects and concepts that make use of nontraditional methods of noise mitigation in the quest for positive soundscape experiences. Both the artists and scientists involved shared the outcomes of these projects, shedding light on how both science and creative endeavour worked in tandem to produce exciting results for the social benefits of applied soundscape solutions for the city.

Following the Sounding Brighton conferences and through extensive work by the Noise Abatement Society and participants, the UK government amended its Noise Policy Statement for England (https://www.gov.uk/government/publications/noise-policy-statement-for-england) as well as its implementation of the EU's Environmental Noise Directive (https://www.gov.uk/government/uploads/system/uploads/attachment_data/file/276239/

noise-action-plan-progress-report-201401.pdf) to include soundscape, and the British Standards Institute (http://www.bsigroup.co.uk/) moved its stance from that of no to neutral to yes, voting for participation in the development of soundscape standards via ISO TC 43/SC 1/Working Group 54 (BS ISO 12913-1:2014). This has subsequently been incorporated into the revision of BS 4142, the UK's most widely used standard for the assessment of environmental noise, to include recognition of soundscape (BS 4142:2014).

10.2.1.10 2013: Soundscape Management in the City's Master Plan for Future Development

Brighton and Hove's Valley Gardens master plan has been developed to restore the aura and impact of the city's once dramatic and beautiful gateway. It forms the main entrance by road from the north, leading down to Brighton's famous seafront. The master plan therefore provides a unique opportunity to raise awareness and promote communication on soundscape management among the general public, stakeholders, and those involved in policy, including encouraging exploration of new ways of listening in local soundscapes, and new ways of tackling noise and improving local soundscape quality. Specifying consideration of a soundscape approach to development within the master plan will position Brighton and Hove City to better understand the effect of soundscapes on community well-being, social cohesion, restorative functions of the environment, and the physical and mental health of individuals, as well as helping it to be at the forefront, globally, of the next generation of urban design, planning, and development.

10.2.1.11 2014: Urban Sound Planning Development (Easteal et al., 2014)

As a result of this commitment by the city (Easteal et al., 2014), and the work of the NAS and EU COST researchers, one of the near-term outcomes has been the inclusion of Brighton and Hove in the company of Rome, Berlin, and Antwerp in the EU-funded SONORUS project "created out of a dream to plan the acoustic environment of our cities in a holistic way" so that "urban sound planning can become an intrinsic component of city planning" (http://www.fp7sonorus.eu/). This revolutionary project will see the training and development of early stage researchers to become the first generation of urban sound planners; this exciting development is building a crucial bridge between architects, planners, and noise control engineers that will lead to the reformation of the development of our towns and cities. In Brighton the work will aid the redevelopment of part of the "spine" of the city, currently an inhibiting "no go" area, into a destination for visitors and citizens (Figure 10.11).

In Brighton and Hove we have seen the powerful and far-reaching results that begin to emerge and can be achieved when a city embraces

Figure 10.11 Valley Gardens, the test site in Brighton and Hove City selected for the SONORUS project. (From Easteal, M., et al., Urban sound planning in Brighton and Hove, presented at Proceedings of Forum Acusticum 2014, 2014, Krakow, Poland.)

a soundscape approach and commits itself to change, and the realization of what R. Murray Schafer (1994:4) so presciently observed nearly four decades ago: "The home territory of soundscape studies will be the middle ground between science, society and the arts."

10.2.2 Soundscape of Waterscapes and Squares on the Sheffield Gold Route

10.2.2.1 Synopsis

Project name: Soundscapes of waterscape and squares on the Sheffield Gold Route

Location: Sheffield, United Kingdom

10.2.2.2 Researchers' Names and Affiliations

Jian Kang: j.kang@sheffield.ac.uk
Yiying Hao: y.hao@sheffield.ac.uk

10.2.2.3 Soundscape Intervention Types

Utilization of noise control elements: The project includes a large stainless steel noise barrier shielding a pedestrian area from a busy road.
Introducing sounds to the soundscape: The project includes a range of innovative water features designed to attract interest and evoke city heritage.

Design alterations: The project was part of an overall improvement of the gateway to the city from the main railway station, including visual improvements and enhanced environments for citizens.

Assessment methodology: Acoustical and psychoacoustical data were extensively analyzed, and citizen questionnaire surveys were carried out.

10.2.2.4 How the Project Has Contributed to Development of Soundscape Concepts

The project represents the most complete study of the sound quality effects of multiple varied waterscapes on citizens' perception and space usage. The interventions each utilize a systematic approach to soundscape measurement and design in a specific local context.

10.2.2.5 Project Overview and Analysis

This example (Kang and Hao, 2011) illustrates the powerful role of soundscapes in shaping cultural heritage and the importance of their inclusion in the regeneration of urban centres and their ability to identify key design parameters. In the case of Sheffield city centre, in the United Kingdom, the primary soundscape of water served to create a central design theme across the city's historic Gold Route, resulting in multiple additional soundscapes as variations stemming from the central concept.

Water soundscapes were chosen because of the element's direct relationship to the city's successful development, including water's core role in the development of the market town in the twelfth century, the steel industry in the fourteenth century, and the industrial revolution in the nineteenth century through to modern times. The field study comprised analysis of the characteristics of multiple types of water features and their effects on listeners using acoustical data, psychoacoustic factors, and field questionnaire surveys.

10.2.2.6 Acoustical Analysis

Spectral analysis, from measurements taken at 1 m from source, showed temporal and frequency variations across a variety of waterscapes along the Gold Route.

The richness and diversity of the waterscapes was also demonstrated through analysis of measurements taken when moving away from the water features. These variations showed the spectral and dynamic ranges of the various soundscapes within a short distance of the source and the design opportunities afforded by these changes, for instance, their potential use in masking traffic noise in public squares (Figure 10.12).

In other cases, the role of water in attracting attention was explored. It was found that as well as the various sounds of water, its visual

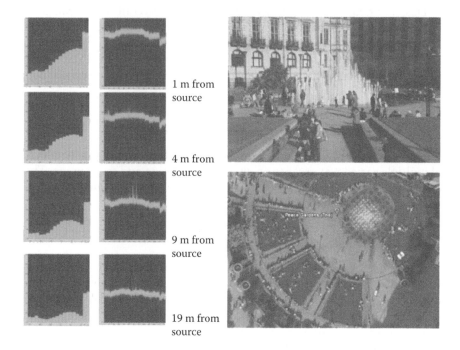

1 m from source

4 m from source

9 m from source

19 m from source

Figure 10.12 Left: Change of soundscape when moving away from the main fountain in the Peace Gardens. Side (top right) and aerial (bottom right) views of the main fountain in the Peace Gardens, Sheffield, UK. (From Kang, J., and Hao, Y., Waterscape and soundscape in Sheffield [Lecture], presented at Meeting of the COST Action TD0804 on Soundscape Examples in Community Context, Brighton, UK, 2011. Copyright © Google.)

effects also played a crucial role in managing the overall soundscape, as demonstrated by the virtually silent installations in Millennium Square (Figure 10.13).

The water features in Sheaf Square provide a concert of activity, creating multitudinous soundscapes varying considerably in spectral and dynamic range (Figure 10.14). The steel barrier erected to reduce noise from the adjacent motorway also serves to create elements of the overall soundscape providing interesting areas of exploration regarding material usage in public squares (Figure 10.14).

10.2.2.7 Psychoacoustic Factors

Analysis of psychoacoustic factors was also made comprising loudness, roughness, sharpness, and fluctuation strength. The results demonstrate the diversity of the soundscapes in terms of Fast Fourier Transform (FFT) versus time. The results, including traffic noise data, highlight the differences with the water sound measurements and are shown in Table 10.2.

Figure 10.13 Virtually silent water features in Millennium Square, Sheffield, UK, attract visitors' attention away from noise disturbance toward a positive soundscape. (From Kang, J., and Hao, Y., Waterscape and soundscape in Sheffield [Lecture], presented at Meeting of the COST Action TD0804 on Soundscape Examples in Community Context, Brighton, UK, 2011.)

Figure 10.14 Left: Cascading and variegated water feature in Sheaf Square, Sheffield, UK. Right: Steel noise barrier with waterfall feature in Sheaf Square, Sheffield, UK. (From Kang, J., and Hao, Y., Waterscape and soundscape in Sheffield [Lecture], presented at Meeting of the COST Action TD0804 on Soundscape Examples in Community Context, Brighton, UK, 2011.)

10.2.2.8 Questionnaire Survey

A series of field questionnaires showed a marked preference for the sounds of the water features in the area, highlighting their noticeability—an interesting finding considering they were not the loudest sounds.

Overall, this soundscape management project demonstrates the importance of utilizing diversity when designing soundscapes in order to create spaces of high cultural value to enhance visitors' enjoyment of an area and reduce noise annoyance.

Table 10.2 Comparison of Psychoacoustic Factors between Different Water Features on the Gold Route

	Psychoacoustic Indices			
	Fluctuation Strength	*Loudness (FFT/ISO 532 B)*	*Roughness*	*Sharpness (FFT/ISO 532 B, Aures)*
Water	*vacil*	*soneGF*	*asper*	*acum*
Barkers Pool	0.0125	21.8	2.46	2.50
Howard Street	0.0198	12.8	2.14	2.91
Peace Gardens	0.00829	30.4	2.80	4.32
Medium cascade	0.0363	24.6	2.70	3.26
Steel barrier	0.0110	19.9	2.45	2.67
Big fountain	0.0111	25.2	2.67	3.29
Small cascade (L1)	0.00956	24.2	2.64	3.29
Small cascade (L3)	0.0189	22.5	2.45	3.17
Traffic noise	0.0112	24.6	2.69	2.17

Source: Kang, J., and Hao, Y., Waterscape and soundscape in Sheffield [Lecture], presented at Meeting of the COST Action TD0804 on Soundscape Examples in Community Context, Brighton, UK, 2011.

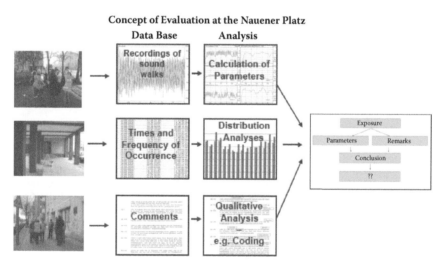

Figure 10.15 The Nauener Platz project in Berlin, Germany, explored understanding and evaluating noise/sound using a holistic approach involving analysis of a variety of qualitative and quantitative parameters. (From Schulte-Fortkamp, B., Soundscape approaches public space perception and enhancement drawing on experience in Berlin [Lecture], presented at Brighton Soundscape Workshop from COST Action TD0804, Brighton and Hove City Council, and the UK Noise Abatement Society, Brighton and Hove, April 6, 2011.)

10.2.3 Nauener Platz: Soundscape Approaches, Public Space Perception and Enhancement Drawing on Experience in Berlin

10.2.3.1 Synopsis

Project name: Nauener Platz: Soundscape Approaches, Public Space Perception and Enhancement, Drawing on Experience in Berlin
Location: Berlin, Germany

10.2.3.2 Researchers' Names and Affiliations

Brigitte Schulte-Fortkamp: b.schulte-fortkamp@tu-berlin.de

10.2.3.3 Soundscape Intervention/Management Type

Utilization of noise control elements: The project includes a physical noise barrier protecting public open space from a busy road.
Introducing sounds to the soundscape: A choice of recorded natural sounds was given to users of special seating facilities within the open space.
Assessment methodology: Local citizens, as the real experts, were extensively involved in initial analysis of the soundscape, in the design of improvements, and in assessing responses, using both quantitative and qualitative methods, including soundwalks, sound measurements, recordings, and interviews.

10.2.3.4 How the Project Has Contributed to Development of Soundscape Concepts

Nauener Platz represents possibly the most complete example (Schulte-Fortkamp, 2011) to date of implementing a systematic approach to soundscape analysis in a specific local context. Those involved in the project have acknowledged that limited resources were available for composition of added sounds, and that giving space users more choice of further developed sounds would be a priority in future evolution of the project.

10.2.3.5 Project Overview and Analysis

The soundscape concept was introduced as a scope to rethink the evaluation of 'noise' and its effects. The challenge was to consider the limits of acoustic measurements and to account for their cultural dimension, introduced by Schafer's neologism and research (Schafer, 1977). Soundscape suggests exploring noise in its complexity and its ambivalence and its approach toward sound to consider the conditions and purposes of its production, perception, and evaluation, to understand evaluation of noise/sound as a holistic approach (Figure 10.15). To discuss the contribution of soundscape research into the area of community noise research means to focus on the meaning of sounds and their implicit assessments to contribute to the understanding that the evaluation through perceptual effects is a key issue.

The development of the Nauener Platz in Berlin is one of the best examples of how to collaborate in a soundscape approach integrating all relevant parties. The project Nauener Platz: Remodelling for Young and Old (Schulte-Fortkamp, 2011) belongs within the framework of the research programme Experimental Housing and Urban Development (ExWoSt) of the Federal Ministry of Transport, Building, and Urban Affairs (BMVBS) by the Federal Office for Building and Regional Planning (BBR). It is related to the fields of research (ExWoSt) concerned with innovation of urban neighbourhoods for families and the elderly. The project executing organization is the Regional Office Berlin-Mitte.

The concept of the development of open spaces relies on the understanding that people living in the chosen environment are the real experts concerning the evaluation of the place according to their expectations and experiences in the respective area. The intention of scientific research in this regard is to learn about the meaning of the noise with respect to people's living situation and to implement adequate procedures to open the 'black box' of people's minds. Therefore, the decision was taken to utilize the combined quantitative and qualitative evaluation procedures inherent within the soundscape approach.

The concept of the development of the Nauener Platz was to rebuild the place into one with social freedom and, from the very beginning, to involve people who live in the area. Therefore, different approaches were taken to get residents involved; for example, public hearings were held about the intentions of the renewing of the place, as well as to get access to the different social groups with respect to their different expectations through well-defined workshops. Also, attention was given to the gender and age of residents, and also to interdisciplinary collaboration.

Such collaboration is considered necessary in the soundscape approach. In this case, it included a collaboration of architects, acoustic engineers, environmental health specialists, psychologists, social scientists, and urban developers. The tasks are related to the local individual needs and were open to noise-sensitive and other vulnerable groups. It was also concerned with cultural aspects and the relevance of natural soundscapes—and related issues of quiet areas, as covered in the European Environmental Noise Directive 2002/49/EC—which is obviously related to the highest level of needs.

The Nauener Platz in Berlin (Figure 10.16) is located between two main roads with a traffic volume of 18,444 cars/24 h on main road 1 and 14,756/24 h on main road 2. The following evaluation procedures were taken: measurements on sound propagation, traffic censuses, binaural measurements, and qualitative evaluations through soundwalks and open interviews introducing the concept of local experts (Figure 10.15).

Figure 10.16 Aerial view of the investigated area of Nauener Platz, Berlin, with measurement positions, determined in consultation with local experts (i.e., residents). The two thin lines are pointing to the lanes of Reinickendorfer Str. (left) and Schulstrasse (right) and were modelled as line sources. (From Schulte-Fortkamp, B., in *Soundscape of European Cities and Landscapes*, ed. J. Kang et al., Oxford: COST Office through Soundscape-COST, 2013, 346. Copyright © Google.)

10.2.3.6 Selection of Local Experts

In a public hearing on the rebuilding of the chosen place, people were introduced to the concept of evaluation through soundwalks; appointments with different groups, differentiated by age and gender, were scheduled with 64 people, respectively. Furthermore, people were informed of the need for narrative interviews. Those interviews were scheduled and conducted with interested people via telephone calls after they participated in the soundwalks.

10.2.3.7 Soundwalks, Measurements, Recordings, Interviews

Each evaluation with soundwalks was carried out in a group of at least eight new experts. The tasks were to define the listening position for the evaluation, rank the road traffic noise using the Rohrmann scale, and comment on those rankings. Noise and comments were binaurally recorded. Based on the recommendations from the local experts while sound walking, the points for subjective evaluation were chosen for comments, rankings, and binaural measurements (Figure 10.16).

10.2.3.8 Improving Local Soundscape Quality

These new approaches and methods (i.e. the soundscape approach) make it possible to learn about the process of perception and evaluation sufficiently, as they take into account the context, ambiance, usual interaction between noise and listener, and multidimensionality of noise perception.

By contrast, conventional methods often reduce the complexity of reality with controllable variables, which supposedly represent the scrutinized object. Furthermore, traditional tests frequently neglect the context dependency of human perception; they only provide artificial realities and diminish the complexity of perception to predetermined values, which do not completely correspond with perceptual authenticity. However, perception and evaluations entirely depend on the respective influences of the acoustic and nonacoustic modifiers in an area.

Following the comments and group discussion, and also the results from the narrative interviews, it could be defined why people preferred some places over the public place and why not. It also became clear how people experienced the noise in the distance from the road, and also with respect to social life and social control. One of the most important findings here is how people react to low-frequency noise in the public place and how experiences and expectations work together. It became obvious that the most-wanted sounds in this area were based on the wishes of the residents to escape the road traffic noise through hearing natural sounds.

10.2.3.9 Reshaping the Place Based on People's Expertise

Relying on the combined evaluation procedures, the place was reshaped, installing a gabion wall along one of the main roads and more audio islands that integrated the sounds people would like to enjoy when using the place (Figure 10.17). While the gabion wall protects against noise around the playground, the newly installed audio islands provide nature sounds as selected by the people involved in the soundscape approach.

10.2.3.10 Conclusions

The process of managing or reducing noise pollution with respect to the expertise of people's minds is related to the strategy of triangulation and provides the theoretical framework with regard to the solutions for changing an area. In other words, approaching the field in this holistic manner is generally needed when noise research is concerned with unknown social worlds.

Figure 10.17 The gabion wall (right) installed in Nauener Platz, Berlin, Germany, to protect park-goers from noise around the playground area; audio islands (left and middle) were also installed to provide areas of respite from traffic noise so that visitors to the park can relax and hear nature sounds. (From Schulte-Fortkamp, B., Soundscape approaches public space perception and enhancement drawing on experience in Berlin [Lecture], presented at Brighton Soundscape Workshop from COST Action TD0804, Brighton and Hove City Council, and the UK Noise Abatement Society, Brighton and Hove, April 6, 2011.)

10.3 OTHER PROJECTS ACHIEVING GOOD SOUNDSCAPE QUALITY

10.3.1 Dublin City Acoustic Planning and Urban Sound Design Programme

10.3.1.1 Synopsis

Project name: Dublin City Acoustic Planning and Urban Sound Design Programme
Location: Dublin, Ireland

10.3.1.2 Researchers' Names and Affiliations

Sven Anderson: http://www.svenanderson.net/

10.3.1.3 Soundscape Intervention Type

Incorporation of sonic art installations: The work of the appointed artist and designer included prototyping sound installations in selected sites.
Assessment methodology: Citizens were engaged through a range of new media.

10.3.1.4 How the Project Has Contributed to Development of Soundscape Concepts

The project has developed a practical repertoire of means through which arts-based practice and urban planning, design, and management can

272 Soundscape and the Built Environment

inspire and mutually support one other. Innovative ways of engaging fresh participants in work on sound have been developed. An ongoing legacy has been created, notably through a *Manual for Acoustic Planning and Urban Sound Design* for the city (Anderson, 2013).

10.3.1.5 Project Overview and Analysis

Dublin City Council (DCC) commissioned artist and designer Sven Anderson to work as the Dublin City acoustic planner/urban sound designer between March 2013 and March 2014. This project was initiated as part of Interacting with the City, part of the Dublin City Public Art Programme (http://www.dublincityarchitects.ie/?p=1011), and is funded through the Department of the Environment, Community and Local Government.

The notion of establishing the role of the urban acoustic planner emerged from the work of Schafer in the 1970s (Schafer, 1977). Several decades later, this idea has developed not only through the continual evolution of the soundscape concept and approach, but also in reaction to fundamental changes of how urban space is planned, mediated, and represented. This project will serve as an interface between a midsized, contemporary urban region (Dublin, Ireland) and a series of research practices and design strategies that place an emphasis on the acoustic dimension of public spaces, civic architectures, and urban experiences.

The project's core methodologies are influenced by the concept of *sonic effects* emanating from CRESSON (*Centre de recherché sur l'espace sonore et l'environnement urbain*) and by diverse projects discussed on the International Ambiances Network (www.ambiances.net), as well as by research outputs emerging from the Soundscape of European Cities and Landscapes (EU COST TD0804, http://soundscape-cost.org/) project. These methods will be extended by observations filtered from contemporary praxis focused on sensory urbanism, experiential architecture, and related technologies. The project will also emphasize the sound installation practices of a select set of sound artists, demonstrating how the methods illustrated through these projects' design and implementation might cross over from the context of contemporary art into the fields of architecture and urban design.

While the project seeks to extend such longer-term research strategies, it is simultaneously focused on developing a concise means of addressing disparate issues in Dublin's urban sound environment without becoming focused on any single issue or project. The project's rapid synthesis of observation, discourse, fieldwork, and prototyping has formed the basis of the *Manual for Acoustic Planning and Urban Sound Design* (http://map.minorarchitecture.org/) (Anderson, 2014), which will inform future project work in Dublin and abroad.

This project commenced in March 2013. The project's outputs include

1. Developing an active project website that serves as a portal for documentation and outreach.
2. Launching a mobile application that provides location-based information related to the project.
3. Reviewing existing noise control infrastructure in the Dublin region.
4. Capturing and cataloguing field recordings of key sonic features in contemporary Dublin, Ireland.
5. Interviewing both local and international project stakeholders and publishing discourse to the project website.
6. Working alongside the planning department to define active research regions within the city.
7. Prototyping sound installations in selected sites.
8. Reviewing these outputs within Dublin City Council in order to provide a reference for similar initiatives in other cities.

By creating this new role within Dublin City Council, and by working with both the planning and public art departments, this project draws attention to the creative potential of the urban sound environment as a medium for design and intervention. Referencing existing noise maps and more traditional noise mitigation techniques that have been implemented in the city (following from the European Environmental Noise Directive 2002/49/EC) as a starting point, this project will focus on considering the introduction of urban sound installations and on promoting discursive design theory stemming from contemporary arts and alternative urban spatial practices within local planning policy.

The work carried out in this project seeks to highlight the richness and complexity of Dublin's urban soundscape as a positive characteristic, which serves to condition the city's unique ambiance and urban experience. The identification of distinct aural typologies within the city, combined with the resources produced through the project's documentation and the construction of the *Manual for Acoustic Planning and Urban Sound Design* (http://map.minorarchitecture.org/), will provide guidelines for protecting, maintaining, and improving local soundscapes that will be integrated in future planning strategies.

10.4 POTENTIAL POLICY APPLICATIONS OF SOUNDSCAPE PRINCIPLES

10.4.1 Soundscape Approach for Early-Stage Urban Planning

10.4.1.1 Synopsis

Project name: Soundscape Approach for Early-Stage Urban Planning
Location: Antwerp, Belgium

10.4.1.2 Researchers' Names and Affiliations

Dick Botteldooren: dick.botteldooren@intec.ugent.be
Bert De Coensel: bert.decoensel@intec.ugent.be

10.4.1.3 Soundscape Intervention Types

Introducing sounds to the soundscape: The project includes encouraging
natural sounds, such as birdsong and moving water.
Design alterations: The project sought to influence layout and design of
a new development from the earliest stage to produce quiet outdoor
spaces with good soundscapes.

10.4.1.4 How the Project Has Contributed to Development of Soundscape Concepts

The work has demonstrated readily replicable methods by which planners
and urban designers can create the potential for good soundscapes in areas
undergoing development (De Coensel et al., 2010). It has also illustrated
means through which soundscapes can be enriched in practical and
self-sustaining ways.

10.4.1.5 Project Overview and Analysis

This project illustrates a possible two-stage applied soundscape approach for
an ongoing redevelopment project in the city of Antwerp, Belgium (De Coensel
and Botteldooren, 2010). The area mainly consists of an abandoned gas
works site and parking lots, and will be redeveloped into a residential area
with room for a coherent public green space. The area is bound by a railway
to the north, a residential area to the south, and major roads to the east.

Tranquil spaces allow urban residents to take respite from periods of sustained directed attention, which characterize modern urban living. There
is growing evidence that having access to quietness close to one's dwelling
may reduce noise annoyance and health effects related to noise.

Long-term measurements (>1 year) in the cities of Antwerp and Ghent
clearly show that inner-city secluded backyards and courtyards provide
a tranquil soundscape, much more than urban parks, which are all too
often located in open areas vulnerable to the intrusion of traffic noise
(Figure 10.18).

Therefore, a good placement of buildings is much more efficient in creating high-quality soundscapes than remediating measures such as placing
noise barriers or absorbing materials.

However, in order to maximize the potential effect of detailed soundscaping measures in urban public spaces, it is essential that aspects of

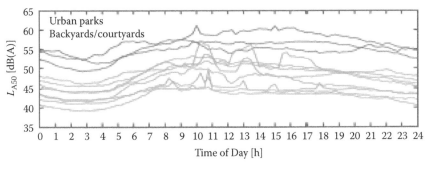

Figure 10.18 Top: Diurnal pattern of LA50 (averages for each 15-min time period of the day, over the complete measurement period) for a number of measurement locations in the cities of Antwerp and Ghent. (From De Coensel, B., and Botteldooren, D., Acoustic design for early stage urban planning, presented at Proceedings of the International Conference on Designing Soundscapes for Sustainable Urban Development, Stockholm, Sweden, 2010.) Bottom: Aerial photograph of the case study area in Antwerp. (Copyright © Google.)

acoustic design should be taken into account at an early stage of the urban planning process.

In the first stage, the arrangement of building blocks, new roads inside the area, squares, and the foreseen urban park was optimized, balancing economic considerations with acoustical objectives, such as limited façade exposure and satisfactory sound levels in the park and courtyards; building placement scenario noise maps were produced (Figure 10.19).

In the second stage, areas of opportunity for soundscape design within the urban park were identified, allowing design measures to be subsequently worked out in detail (e.g., installing appropriate water features or adding greenery that is attractive to particular songbirds). Detailed propagation

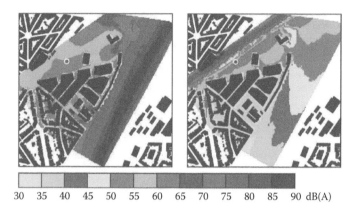

30   35   40   45   50   55   60   65   70   75   80   85   90 dB(A)

Figure 10.19 Noise maps (LAeq, day) for road traffic (left) and railway traffic (right) for a particular building placement scenario. The blue dot represents a particular listener location inside the urban park. (From De Coensel, B., and Botteldooren, D., Acoustic design for early stage urban planning, presented at Proceedings of the International Conference on Designing Soundscapes for Sustainable Urban Development, Stockholm, Sweden, 2010.)

simulations combined with virtual listening tests that use models of auditory perception could be employed at this stage (Figure 10.20). The approach can be used to draw masking maps, delimiting the area for which design elements are effective in drawing attention away from unwanted sound.

10.4.2 Soundscape-Based Criteria Used in Territorial Planning: A Case in the Region of the Alpujarra, Granada, Spain

10.4.2.1 Synopsis

Project name: Soundscape-Based Criteria Used in Territorial Planning: A Case in the Region of the Alpujarra, Granada, Spain
Location: Granada, Spain

10.4.2.2 Researchers' Names and Affiliations

Antonio Torija: ajtorija@ugr.es
Diego P. Ruiz: druiz@ugr.es

10.4.2.3 Soundscape Intervention Type

Utilization of sounds that already exist in the location: Soundscape quality was assessed to determine needs for protection and recovery, and where activities with potential implications for the sound environment might be hosted.

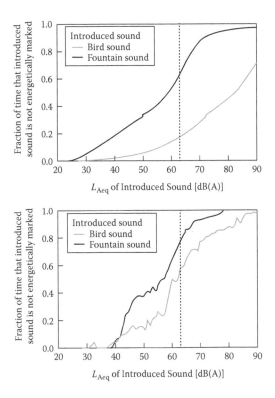

Figure 10.20 Fraction of time that fountain or bird sound is audible (left) and attracts attention (right) for different sound levels. The dashed line marks the sound level of the traffic at the considered location. (From De Coensel, B., and Botteldooren, D., Acoustic design for early stage urban planning, presented at Proceedings of the International Conference on Designing Soundscapes for Sustainable Urban Development, Stockholm, Sweden, 2010.)

Assessment methodology: A Soundscape Quality Index (SQI) was developed, derived from analysis of citizen perceptions.

10.4.2.4 How the Project Has Contributed to Development of Soundscape Concepts

This project has demonstrated a practical way in which soundscape criteria can be represented in land management and planning at the regional scale, including through development of a Soundscape Quality Index (Torija and Ruiz, 2013).

10.4.2.5 Project Overview and Analysis

The University of Granada has developed a methodology for evaluating the quality of soundscapes, with the final objective of including sound criteria

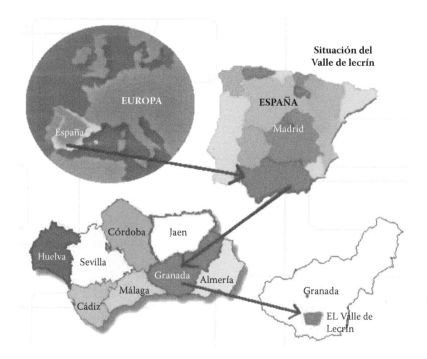

Figure 10.21 Location of the County of Valle de Lecrín in the province of Granada, Spain. (From Torija, A., and Ruiz, D.P., in *Soundscape of European Cities and Landscapes*, ed. J. Kang et al., Oxford: COST Office through Soundscape-COST, 2013, 352. Copyright © Google.)

in territorial planning and land management. This work was framed in a collaboration agreement between the University of Granada and the Provincial Authority of Granada. The developed methodology was applied in the County of Valle de Lecrín (Figure 10.21) in the province of Granada, Spain (Torija and Ruiz, 2013).

It was considered that soundscape criteria needed to be incorporated in land management and territorial planning because of the strong impact of soundscape on the welfare of the population who interact with it. This would also allow the protection and conservation of soundscapes of special interest. Incorporating soundscape criteria in territorial planning involved assessing soundscape quality, to determine which soundscapes need some protection, which should be recovered, and which could host an activity with a potential significant impact on the sound environment.

To evaluate the quality of the soundscapes of the county, an acoustical assessment was conducted. To accomplish this, a set of descriptors to characterize the loudness, temporal structure, and spectral composition were calculated.

Before undertaking the field measurement campaign, analysis of the study area was carried out. This included grouping the county into different subareas according to the type of predominant sound source, presence of noisy sources (urban, roads, industry, etc.), presence of water fountains or watercourses, vegetation (type, density), presence of birdlife, topography, and land use. Having identified these heterogeneous areas, a random selection of measurement points within each of the previous areas was used, so the whole typology of soundscapes in Valle de Lecrin was covered. A subjective evaluation was also conducted, using surveys of people's perceptions of each of the soundscapes.

For evaluating and characterizing soundscape quality, the indicator Soundscape Quality Index was developed (Torija and Ruiz, 2013). This indicator assessed the quality of a soundscape on the basis of its degree of naturalness (presence of natural sound sources such as birds, vegetation, and watercourse) and absence of human-generated sources of noise (traffic, industrial activities, commercial/recreational activities, etc.), but also in terms of people's perception.

A set of sound descriptors was developed with the goal of identifying and quantifying the impact of traffic-related and mechanical sounds, and also the presence of natural sound sources (birdlife and watercourses). The SQI value was given by the value of each of the sound descriptors weighted by a series of correction coefficients obtained from perceptual assessment done by the surveyed population (Torija and Ruiz, 2013).

Five soundscape categories were defined and mapped (Figure 10.22):

Category C1: Degraded sound environment
Category C2: Environment with low sound importance
Category C3: Environment with sound importance
Category C4: Environment with high sound importance
Category C5: Environment with special sound importance

The methodology enabled diagnosis of the quality of a given soundscape, giving information relevant for management. It also informed action planning oriented toward conservation, protection, or restoration of a given soundscape, including identification of less sound impacting alternatives for hosting certain activities and prioritizing noise source control.

10.4.3 Evaluating Indoor Soundscapes by Questionnaires

10.4.3.1 Synopsis

Project name: Evaluating Indoor Soundscapes by Questionnaires
Location: Sheffield, United Kingdom

10.4.3.2 Researchers' Names and Affiliations

Papatya Nur Dökmeci: p.dokmeci@sheffield.ac.uk
Jian Kang: j.kang@sheffield.ac.uk

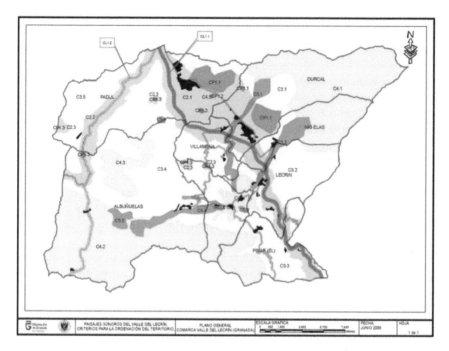

Figure 10.22 Soundscape management map, based on Sound Quality Index (SQI) indicators developed by the University of Granada, showing categorization values ranging from C1 to C5 for the County of Valle de Lecrín in the province of Granada, Spain. (From Torija, A., and Ruiz, D.P., in *Soundscape of European Cities and Landscapes*, ed. J. Kang et al., Oxford: COST Office through Soundscape-COST, 2013, 352. Copyright © Google.)

10.4.3.3 Soundscape Intervention Type

Assessment methodology: Questionnaire surveys were carried out for users of three libraries, in conjunction with acoustic measurements.

10.4.3.4 How the Project Has Contributed to Development of Soundscape Concepts

This work (Dökmeci and Kang, 2013) showed how techniques hitherto used mainly in open-air situations and at city scale could be used to assess users' perceptions of the quality of indoor spaces. The study yielded valuable lessons for future study of indoor soundscapes.

10.4.3.5 Project Overview and Analysis

Soundscape studies have hitherto concentrated mainly on open-air soundscapes and the city scale. The aim of this study was to present an example

	Information Commons	Western Bank	ST. George's
Foyer location	1st floor	1st floor	**Entrance**
Area	**372 m^2**	368 m^2	362 m^2
Volume	**2,667 m^3**	1,945 m^3	1,548 m^3
Plan	**L-shape**	Rectangular	Rectangular
Atrium location	Above h: 14 m	**Below h: 4 m**	Above h: 10.5 m
Atrium void	**1,638 m^3**	105.5 m^3	424 m^3
Skylight	Above atrium	Above atrium	Above atrium
Wall material	Plaster & wooden acoustic panels	Wood panels & glass sheets	Painted cinder block
Ceiling material	Concrete & plaster acoustic panels	Hard semi-transparent plastic	Painted concrete
Floor material	Carpet	Vinyl	Carpet
Crowd level	**40-Pass, 65-Still**	22-Pass, 14-Still	25-Pass, 10-Still

Figure 10.23 Physical characteristics of the foyers of three libraries in Sheffield, UK, chosen for a study to evaluate indoor soundscape quality. (From Dökmeci, P., and Kang, J., in *Soundscape of European Cities and Landscapes*, ed. J. Kang et al., Oxford: COST Office through Soundscape-COST, 2013, 345.)

of subjective acoustic assessment tools that could be integrated with soundscape studies for indoor environments. The key part of designing indoor soundscape questionnaires is to understand how the users perceive the soundscape in an identified context within an enclosure (Dökmeci and Kang, 2013).

Three different libraries in Sheffield, UK, used mainly by university students, were chosen for this study: Western Bank, Information Commons, and St. George's. The main foyer areas in each library were used for sound measurements and recordings, carried out simultaneously with questionnaire surveys in order to achieve reliable comparisons and correlations. The library foyers varied in spatial characteristics, acoustic absorbency, equipment, and other features (Figure 10.23).

The questionnaire comprised four parts:

1. Open-ended questions on usage of the space
2. Time spent, like or dislike, basic demographics
3. Five-point rating scale for the evaluation of different factors in the space
4. Five-point rating scale for the subjective evaluation of sound sources

No.	Correlated Factors (Quality)	Disturbance/preference of the overall noise		Annoyance caused by the overall noise	
		Spearman's Rho Test			
		r	p	r	p
1	Level of acoustic comfort	.346	p<0.01	−.306	p<0.01
2	Intelligibility of sounds	.334	p<0.01	−.218	p<0.01
3	Ability to locate via sounds	.222	p<0.01	−.282	p<0.01
4	Level of sounds	.226	p<0.05	−.253	p<0.05
5	Way-finding	.219	p<0.05	−.267	p<0.05

Figure 10.24 Questionnaire comprising correlated sound quality factors used to evaluate the indoor soundscapes of the foyers of three libraries in Sheffield, UK. (From Dökmeci, P., and Kang, J., in *Soundscape of European Cities and Landscapes*, ed. J. Kang et al., Oxford: COST Office through Soundscape-COST, 2013, 345.)

The questionnaire included factors related to indoor physical comfort (acoustics, air quality, humidity, temperature, light), acoustics (sound level, sound types, sound intelligibility, reverberation level, noise from other spaces, locating by sound), and architecture (way finding, spaciousness, crowding) (Figure 10.24). Initial pilot studies were used to identify the 19 different sound sources included in the questionnaire for evaluation. The most and least annoying sound sources were identified for further investigation. A simple random sampling technique was used to secure 30 participants in each of the three libraries.

Objective parameters, including sound pressure level, frequency spectrum, loudness, roughness, and sharpness, were measured using a Neumann KU100 dummy head with Edirol R-44 portable recorder. Audio samples were analyzed using ArtemiS psychoacoustic analysis software.

The surveys showed that (Figure 10.25):

1. Architectural factors play an important role in soundscape evaluation.
2. In all three libraries, way finding, thermal comfort, and air quality were found to have the highest qualities, and the least quality was rated for spaciousness and crowd.

	Factorial Importance	Factorial Quality	Noise Annoyance	Sound Preference
Female 	−Level of thermal comfort (p<0.05)	−Level of reverberation (p<0.05)	−Mobile phones (p=0.078) −Personal music players (p=0.075)	−
Male 	−Ability to locate via sounds (p<0.05) −Noise from neighbouring spaces (p<0.05)	−	−	−Laughter (p<0.05)

Figure 10.25 Significant effects were detected when correlations among gender, academic level, and usage variances were analyzed for factor ratings and annoyance from sound sources in the questionnaire study evaluating the indoor soundscapes of the foyers of three libraries in Sheffield, UK. (From Dökmeci, P., and Kang, J., in *Soundscape of European Cities and Landscapes*, ed. J. Kang et al., Oxford: COST Office through Soundscape-COST, 2013, 345.)

3. In all three libraries, sound sources, including mobile phones, personal music players, and construction noise, were found to be the most annoying, and the least annoying noted as walking/footsteps and page turning.
4. Correlations among factors were analyzed to understand cross-effects of different factors, especially with acoustic comfort.
5. Significant effects were detected when correlations among gender, academic level, and usage variances were analyzed for factor ratings and annoyance from sound sources.
6. There were considerable differences in objective parameters among the survey sites.
7. Correlation analysis showed that sound pressure level and loudness were significantly related with subjective evaluation.

The study indicated that further assessment should be carried out by detailed architectural and objective postsignal analysis. Evaluation should include questionnaires designed for specific purposes. A multidimensional and holistic approach was needed for design of indoor soundscape studies.

10.4.4 Soundwalks for the Urban Development of Natural Spaces in Alachua County, Florida

10.4.4.1 Synopsis

Project name: Soundwalks for the Urban Development of Natural Spaces in Alachua County, Florida
Location: Florida, United States

10.4.4.2 Researchers' Names and Affiliations

Jose A. Garrido: jagarrido@ugr.es
Gary W. Siebein: gsiebein@siebeinacoustic.com
Diego P. Ruiz: druiz@ugr.es

10.4.4.3 Soundscape Intervention Type

Assessment methodology: An innovatory analytical soundwalk method was developed, capable of use by nonacousticians.

10.4.4.4 How the Project Has Contributed to Development of Soundscape Concepts

This study (Garrido et al., 2013) developed a soundwalk method to characterize acoustic rooms accessible to architects, urban planners, and others without detailed expertise in acoustics. The method enabled key acoustic features to be identified, and provided guidance to inform design of potential development.

10.4.4.5 Project Overview and Analysis

In Alachua County, Florida, a revised soundwalk method was developed. As opposed to static observations at predetermined locations, the soundwalk, intended as an exploratory tool for the detection of gradual acoustic changes over time and space within large environments, allowed the route to be readjusted on the basis of attention-grabbing elements spotted on the way, thus yielding a more accurate definition of acoustic rooms and other soundscape elements (Garrido et al., 2013).

The emphasis in this kind of soundwalk was on subtle variations of acoustic conditions that could only be appreciated on foot, rather than using a motorized vehicle on a preset itinerary of points. Individual soundwalks would promote an immersive and more discriminating listening experience. The method also aimed to be accessible to architects and urban planners lacking advanced knowledge of acoustics.

An 8 h (1100–1900) 8-mile individual soundwalk was performed in a natural area, including woodlands, cattle farms, and a wetland corridor. The route wound along narrow county roads through the municipalities of Hawthorne,

Figure 10.26 Aerial photo of timberlands in Alachua County, Florida, showing an 8-mile soundwalk route and regular stopping points (marked with gray spots) and outstanding acoustic rooms (marked with black spots). The wetland corridor is shaded in white. (From Garrido, J.A., et al., in *Soundscape of European Cities and Landscapes*, ed. J. Kang et al., Oxford: COST Office through Soundscape-COST, 2013, 354. Copyright © Google.)

Campville, and Windsor. Audio and video recordings of characteristic sonic events, panoramic pictures of acoustic rooms, and brief informative notes of interesting facts were thoroughly taken (Figure 10.26). An acoustic room was defined as a finite space whose physical features create a characteristic sound field in accordance with the theory of sound propagation.

Sound sources detectable to the human ear at selected locations were listed, and further information on temporal patterns of occurrence and spatial distribution was collected. Initial assessment of sonic events in terms of key perceptual properties such as pitch was attempted. These data were gathered, processed, and broken down into schematic taxonomies referred to specific time periods for every stopping point on the route. In addition, commonalities and differences, balances and imbalances resulting from competing sources, and other interesting particularities could be unveiled from the information collected in the field.

Figure 10.27 The soundwalk study conducted in Alachua County, Florida, identified areas of special acoustic interest for priority designation for designers of any new development. (From Garrido, J.A., et al., in *Soundscape of European Cities and Landscapes*, ed. J. Kang et al., Oxford: COST Office through Soundscape-COST, 2013, 354.)

The study identified the well-rounded canopy structure of the natural wetland corridor as one of the area's most outstanding acoustic features that should be considered a priority for designers of any new development (Figure 10.27). The canopy, defined as the uppermost spreading branchy layer of a forest formed by mature tree crowns, provides shelter from predators and unfavourable weather conditions for a wide range of arboreal animals and other organisms. In particular, bird calls and songs of numerous species, coming from the upper parts of the trees, show up as keynotes of this sonic environment.

However, it appeared that conservation of the special sonic environment created by the canopy did not necessarily require absolutely unspoiled nature. Habitation strategies used by the local rural community, such as single-story, open-plan houses with low-pitched roofs, and recycled shipping containers and railroad cars, did not seem to have unduly disrupted the canopy soundscape of birdsong.

10.4.5 Approach to the Evaluation of Urban Acoustical Comfort Using the Soundscape Concept

10.4.5.1 Synopsis

Project name: Approach to the Evaluation of Urban Acoustical Comfort Using the Soundscape Concept
Location: Istanbul, Turkey

10.4.5.2 Researchers' Names and Affiliations

Asli Ozcevik: asliozcevik@hotmail.com
Zerhan Yuksel Can: karabi@yildiz.edu.tr

10.4.5.3 Soundscape Intervention Type

Assessment methodology: Sound measurements and recordings taken on soundwalks were analyzed to calculate sound quality metrics, and surveys of subjective perceptions were carried out, defining the particular contribution of soundmarks to soundscape identity.

10.4.5.4 How the Project Has Contributed
to Development of Soundscape Concepts

This project (Ozcevik and Can, 2013) demonstrated a practical approach for evaluating, conserving, and rehabilitating acoustical comfort in urban areas, with a particular focus on the role of soundmarks in defining local identity, along with suggestions for how future environments might be improved.

10.4.5.5 Project Overview and Analysis

The doctoral thesis "An Approach on the Evaluation of Urban Acoustical Comfort with the Soundscape Concept" (Ozcevik and Can, 2013) aimed to develop an applied soundscape approach for the evaluation, conservation, and rehabilitation of acoustical comfort in urban areas by the data obtained from field and laboratory studies, allowing the correlation of subjective and objective information. It tested the hypothesis that soundscape quality can objectively be evaluated through the perceptibility of soundmarks.

In this approach, in situ measurements (sound level measurements, binaural sound recordings) and sound quality metrics are defined as the objective data, and pairs of adjectives suitable for describing the sound environment (obtained by semantic differential test), surveys (questionnaire, semantic differential test), jury, and listening tests are obtained as the subjective data.

The approach was revised after first being tested theoretically using initial locations in Istanbul, Turkey (Besiktas Pier Square, Ortakoy Pier Square, Bagdat Street, and Barbaros Boulevard), and then practically using new locations (Yildiz Park, Bebek Park, Meclis-i Mebusan Street and Findikli Park, and Serencebey Park).

The methodology used in this applied soundscape approach was

1. Area selection—at least three areas because of two unknown quantities in listed equations—and analysis of the sound environments (i.e., description of soundscapes, determination of soundmarks, and previsions of acoustical satisfaction, realized by the acoustic expert)
2. In situ measurements using soundwalks—sound measurements and binaural sound recordings, lasting 15 min, and realized by considering the proper conditions for quantity and quality of the recordings

3. Preparation and analyses of the recordings—calculation of sound quality metrics, obtaining edited 5 min periods having only the predicted soundmarks and nine sound segments, each of 3 min, calculating statistical values of the sound quality metrics for the quantitative confirmation of the edited period
4. Use of listed equations for pairs of adjectives suitable for describing the sound environment (obtained by semantic differential test)
5. Interpretation of the results in a holistic way—overall and detailed evaluations of the sound environments using radar graphics

The study found that there was a direct relationship between the satisfaction from soundmark and the sound environment, and therefore sound environment could be assessed depending on the soundmark (Ozcevik and Can, 2013). The proposed applied soundscape approach allowed distinguishing unpleasant sounds from preferred sounds peculiar to the area worked on, thus facilitating decisions to improve the sound environment. Visual presentation using radar graphics provided a holistic image of the acoustical comfort degree, as well as comparison of different areas.

Soundmarks create the acoustical identity of an environment. The culture, society, city, and zone all together shape the soundmarks. Trying to fight noise disregarding soundmarks might harm acoustic identity and, in some cases, acoustical heritage. Once desirable and undesirable soundmarks are discriminated, it becomes possible to manipulate the sound environment by strengthening perception of desirable soundmarks, including by masking undesirable sounds. Local soundscape quality of cultural heritage zones may also be conserved or rehabilitated, depending on the specific soundmarks.

10.5 EXAMPLES OF SOUNDSCAPE PRACTICE IN POLICY

10.5.1 Sound Islands: Bilbao Municipality Policy to Improve Citizens' Quality of Life through Soundscape Design and Management

10.5.1.1 Synopsis

Project name: Sound Islands: Bilbao Municipality Policy to Improve Citizens' Quality of Life through Soundscape
Location: Bilbao, Spain

10.5.1.2 Researchers' Names and Affiliations

Igone Garcia Perez: igone.garcia@tecnalia.com
Itziar Aspuru Soloaga: itziar.aspuru@tecnalia.com

10.5.1.3 Soundscape Intervention/Management Type

Utilization of noise control elements: Including a noise barrier and more green landscaping.

Introducing sounds to the soundscape: Including a water feature with varying flow patterns.

Design alterations: Including reducing speed, transport rerouting, and new facilities.

Assessment methodology: An indicator of environmental sound experience quality was developed using both sound measurements (events and sources, as well as level) and subjective evaluation of sound sources and their congruency.

10.5.1.4 How the Project Has Contributed to Development of Soundscape Concepts

This project (García Pérez and Aspuru Soloaga, 2013) included development of a practical methodology, incorporating application of an environmental sound experience quality indicator, to assess specific places using both measurements and questionnaire surveys. This analysis also supported identification of specific actions to improve soundscapes.

10.5.1.5 Project Overview and Analysis

The aim of the project—municipality resources and considering the framework of 0LIFE10/ENV/IT/407: QUADMAP (Quiet Areas Definition and Management in Action Plans) http://www.quadmap.eu/—is to create a network of sound islands in Bilbao municipality. Bilbao has defined the concept of sound island as an urban public space with a soundscape that invites citizens to relax. The Bilbao municipality goal is to create at least one sound island per district (eight total).

This network will integrate sound islands that already exist in Bilbao and other places that will become sound islands with some interventions. Considering this second approach, the first sound island in Bilbao that will be declared is General Latorre Square. This square underwent comprehensive renovation with an urban design that includes soundscape as a variable to define the interventions (Figure 10.28).

To integrate the soundscape approach in the urban design criteria in General Latorre, a specific indicator Environmental Sound Experience Indicator (ESEI) was applied (García Pérez et al., 2012). This indicator was developed by Tecnalia, and it assesses the environmental sound experience quality. The ESEI indicator considers not only sound levels, but also the number of events (that are relevant with respect to the background level) and the composition of the sounds sources in the area.

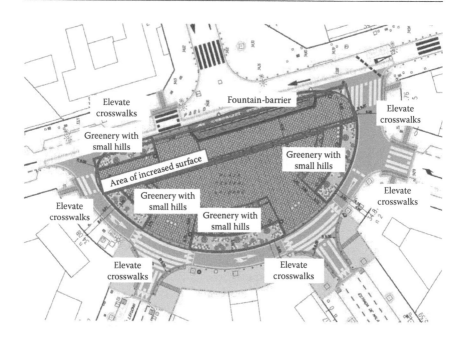

Figure 10.28 Design of the General Latorre, Bilbao Square, with the designated actions to improve the acoustic comfort of the space as perceived by citizens. (From García Pérez, I., and Aspuru Soloaga, I., in *Soundscape of European Cities and Landscapes*, ed. J. Kang et al., Oxford: COST Office through Soundscape-COST, 2013, 350.)

To calculate this indicator, not only sound measurements must be developed, but also a psychosocial analysis (with questionnaires) is required to include the subjective evaluation of sound sources: it relates to whether the sources that characterize the sound atmosphere are perceived as pleasant or unpleasant and if they are congruent (expected to be listened to) in the specific place.

In General Latorre Square, sound measurement and questionnaires were developed simultaneously in the moments of the day when the area is used by citizens (García Pérez and Aspuru Soloaga, 2013). The results of the study show a value of ESEI of 5.2 on a 0 to 12 scale. To improve the soundscape (increase the value of ESEI), some actions must be addressed to reduce noise levels (related to traffic) as a minimum requirement and improve soundscape (in connection with the results of the perception analysis).

Bilbao municipality considers that soundscape is one of the principal variables to increase the acoustic comfort in urban public spaces. There are some experiences (Herranz-Pascual et al., 2011) that show some evidence of the restorative capacity from stress of positive soundscapes in urban areas. These concerns of the city council were the starting point of the sound island policy.

A sound island of Bilbao must meet certain requirements of ESEI indicator, sound level, citizen satisfaction with the area and its soundscape, and uses and functions of the spaces (an area for relaxing activities). After the intervention in General Latorre, it is expected that the area will improve its soundscape and meet these requirements (in comparison with the previous scenario).

10.5.1.6 Applied Soundscape Approach

To integrate a soundscape approach in the design of the General Latorre intervention, some sonic challenges were defined to give answer to the different variables that must be addressed to increase the ESEI results (and acoustic comfort). The first challenge was referred to traffic noise pollution (considering not only the level, but also the events), as it was reported by citizens as one of the principal factors for acoustic discomfort in the area. The second challenge was in connection with increasing the presence of children (less than 3 years old) and other natural sounds and the related events.

To give answer to these challenges, all the results of the perception analysis were considered (Table 10.3). This analysis gives information about what actions will be better perceived by citizens considering a holistic approach.

In the General Latorre example, the following specific actions were included (Table 10.3):

- To reduce traffic noise, there will be actions to reduce speed and deter private cars' passage through the square (elevated crosswalks in all the pedestrian accesses to the square will be created).
- To increase the surface for resting and being, the pass-by area will be included in the square.
- To modify the acoustic atmosphere, a water fountain was designed. This fountain will also be a noise barrier that protects the square from the principal traffic source in the area (the most used street with traffic jams in rush hour). This fountain will have a ripple waterfall (with rolling stones), and it will also have sound events created with water jets adjustable in intensity and duration.
- To modify the acoustic atmosphere, more greenery will be also put in the area (using trees, shrubs, and grass), with small hills to create new diffraction borders.

The experience in General Latorre to integrate soundscape as a criteria for urban public spaces design will be applied in other public place renovation processes in Bilbao. It can also be considered and applied in other European cases. This experience can be summarized in four steps (García Pérez and Aspuru Soloaga, 2013):

Table 10.3 Analysis of the Consequences in ESEI of Different Possible Actions
Defined to Give Answers to the Sonic Challenges in General Latorre
Square, Bilbao

Actuation	Approach	LAeq (dBA)	Negative Events (%)	Positive Events (%)	ESEI (new)
Actual Situation		63			5–5.2
A1: Closing a street	Conservative	−2	−30	20	6.2–6.3
	Possibilitative	−4	−70	40	7.5–7.5
A2: Pedestrian preference	Conservative	−0.5	−20	—	5.4–5.6
	Possibilitative	−1	−40	—	5.8–6.0
A3: Traffic fluency	Conservative	1	−20	−10	5.1–5.3
	Possibilitative	−1	−40	10	5.9–6.0
A4: Barrier	Conservative	−1.5	−10	—	5.5–5.7
	Possibilitative	−2.5	−20	10	5.9–6.1
A5: Absorption	Conservative	0	—	—	5–5.2
	Possibilitative	−0.4	—	—	5.1–5.4
A7: Vegetal elements	Conservative	0	—	20	5.3–5.6
	Possibilitative	0	—	50	5.3–5.6
A8: Water elements	Conservative	0	—	20	5.3–5.6
	Possibilitative	+3*	—	20	5.6–5.9
A9: Children	Conservative	0	—	20	5.3–5.6
	Possibilitative	+3*	—	50	5.7–5.9

Source: García Pérez, I., and Aspuru Soloaga, I., in *Soundscape of European Cities and Landscapes*, ed. J. Kang et al., Oxford: COST Office through Soundscape-COST, 2013, 350.

Note: Two different scenarios were defined in order to facilitate the decision-making process.

Step 1: Assessment of the acoustic comfort or soundscape approach in the preoperational scenario.

Step 2: Conclusions of the analysis and definition of sonic challenges in the area.

Step 3: Ideas for the intervention (to give answer to the sonic challenges). In this step, the active participation of the architects of the area is required.

Step 4: Integration of the specific ideas in the project and improvement of the actions from a soundscape point of view.

10.5.2 Enhancing Soundscapes through Sustainable Urban Design, Transport, and Urban Greening in Vitoria-Gasteiz

10.5.2.1 Synopsis

Project name: Enhancing Soundscapes through Sustainable Urban Design, Transport, and Urban Greening in Vitoria-Gasteiz

Location: Vitoria-Gasteiz, Spain

10.5.2.2 Researchers' Names and Affiliations

Jose Luis Bento Coelho: bcoelho@ist.utl.pt
Alberto Bañuelos: abi@aacacustica.com

10.5.2.3 Soundscape Intervention/Management Type

Utilization of noise control elements: In the creation of a new mobility structure based on superblocks.

Introducing sounds to the soundscape: Promoting green spaces in the city to increase biodiversity with more natural sounds, especially of birds, and through river recovery.

Design alterations: Including reducing use of motorized vehicles, making areas inaccessible to traffic, and creating more pedestrian areas.

10.5.2.4 How the Project Has Contributed to Development of Soundscape Concepts

This work (Bento Coelho and Bañuelos, 2013) demonstrated how soundscape considerations could be integrated within a holistic area-wide planning and transportation approach. The project showed how local soundscapes could be characterized within an action plan context.

10.5.2.5 Project Overview and Analysis

The future of cities depends on approaches that ensure sustainability. This needs new urban structures and actions promoting lower energy consumption, respect for the environment, and health. Urban mobility must be transformed from overreliance on private vehicles toward public transport and nonmotorized mobility, allowing more pedestrian areas and more enjoyment of public space. Reducing traffic noise gives the opportunity to incorporate soundscape quality in urban design, creating more attractive public spaces and encouraging more use of cycle and walking routes.

The example of Vitoria-Gasteiz in Spain (Bento Coelho and Bañuelos, 2013), European Green Capital 2012, includes the creation of a new mobility structure based on the superblock concept, creating a new kind of public space with greater liveability, where the pedestrian regains prominence in the city. Moving from conventional urban noise control to creative soundscape design will improve health and quality of life.

Traffic noise makes a marked contribution to urban pollution. Reducing use of motorized vehicles, making areas inaccessible to traffic, and urban redesign aimed at achieving more pedestrian areas accessible for public enjoyment imply a reduction in environmental noise, opening opportunities to incorporate soundscape design in the future city. An urban soundscape approach can help create more pleasant recovered spaces for public use,

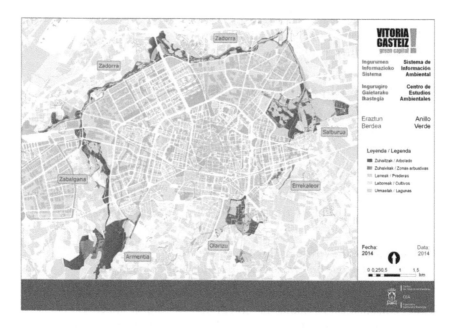

Figure 10.29 The historic green belt surrounding Vitoria-Gasteiz comprises a series of "peri-urban parks of high ecological and landscape value, strategically linked by eco-recreational corridors." (From Vitoria-Gasteiz City Council, The greenbelt of Vitoria-Gasteiz, 2015, http://www.vitoria-gasteiz.org/. Copyright © Vitoria-Gasteiz City Council.)

including creation of pleasing sound environments that promote the use of bicycle and pedestrian pathways.

Promoting green spaces in the city to increase biodiversity helps change the sound environment. Urban transformation both reduces traffic noise and encourages natural sounds within the city. Changing the city soundscape, including new urban sounds, will transform people's sound perception of the modified spaces.

The green belt of Vitoria-Gasteiz has been incorporating natural spaces into suburban areas, bringing natural sounds closer to the population (Figure 10.29). A new phase of increasing biodiversity in the city is underway with the creation of an inner green belt. This aims to transform the urban landscape and soundscape. Less motorized traffic will reduce ambient noise levels in streets that are part of the inner belt. More natural sounds will be encouraged due to the increased presence of wildlife, especially birds. The soundscape will also be improved by other sources, such as water, for example, in recovery of the River Batán at Avenida Gasteiz, the first stage of this project.

This process is part of the Sound Environment Enhancement Action Plan of the municipality of Vitoria-Gasteiz (Vitoria-Gasteiz City Council, 2012).

The plan's preliminary stages have already taken account of soundscape, with characterization plans giving a reference to assess the changes that initial actions will make.

10.6 PRIORITIES FOR SOUNDSCAPE INTERVENTION

10.6.1 Intangible Cultural Heritage Value of Unique and Intrinsic Soundscapes and Soundmarks

10.6.1.1 Synopsis

Project name: The cultural value of the soundscape of Folk Festivals
Location: Italy

10.6.1.2 Researchers' Names and Affiliations

Luigi Maffei: Luigi.MAFFEI@unina2.it
Giovanni Brambilla: giovanni.BRAMBILLA@idasc.cnr.it
Maria Di Gabriele: Maria.DIGABRIELE@unina2.it

10.6.1.3 Soundscape Intervention/Management Type

Utilization of sounds that already exist in the location: In the context of assessing intangible cultural heritage, the project aimed to identify if rhythms imposed by music or voice represent the soundmarks of some folk festivals.

Assessment methodology: This included historical analysis, based on bibliographic research and interviewing local experts, combined with in situ audio binaural and video recordings, and in-field questionnaires to assess citizen perceptions (Figure 10.30).

10.6.1.4 How the Project Has Contributed to Development of Soundscape Concepts

The study represents the most complete analysis of the effects and identification of soundscapes as intrinsic, intangible cultural heritage assets (Maffei and Di Gabriele, 2013; see also Chapter 9 for the full project description and details of the work). Its findings are aligned with UNESCO's international Convention for the Safeguarding of the Intangible Cultural Heritage (2003).

10.6.1.5 Project Overview and Analysis

With the international Convention for the Safeguarding of the Intangible Cultural Heritage, UNESCO recognizes the "intangible" as "cultural

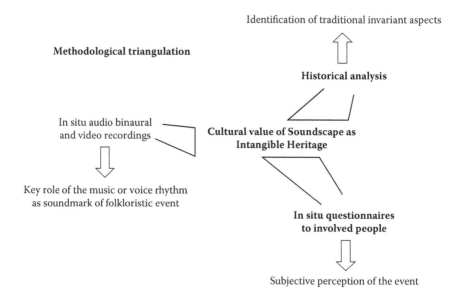

Identification of traditional invariant aspects

Methodological triangulation

Historical analysis

In situ audio binaural
and video recordings

Cultural value of Soundscape as
Intangible Heritage

Key role of the music or voice rhythm
as soundmark of folkloristic event

In situ questionnaires
to involved people

Subjective perception of the event

Figure 10.30 Triangulation methodology applied for the identification of the cultural value of soundscape of folk festivals (Maffei and Di Gabriele, 2013; see also Chapter 9 for a detailed description of the study). (From Maffei, L., and Di Gabriele, M., in *Soundscape of European Cities and Landscapes*, ed. J. Kang et al., Oxford: COST Office through Soundscape-COST, 2013, 355.)

heritage" (UNESCO, 2003). The main criteria for inscription of these properties on the UNESCO representative list are their representativeness of human diversity and creativity. So far, the nomination and selection of an intangible property for inclusion in the list do not require any acoustic criteria, and the "sound" is not considered an intrinsic value of cultural heritage that contributes to perceptual experience of the event. Up to now, although some studies aimed at identifying songs and dances as sound-marks for local people were made by the Japanese scientific community (Maffei and Hiramatsu, 2006), no specific research has been conducted on the soundscape of folk festivals, although they involve a strong participation of population.

The soundscape of folk festivals, in fact, can also be considered a part of intangible cultural heritage, if it is unique or possesses qualities that make them recognizable by the people and attractive to tourists (Maffei, 2011; see also Chapter 9 of this book for the full project description and details). This concept, also applied to soundmarks and unique and intrinsic sound-scapes in general, is further reinforced by Böhme's premise that our sense of "home" is directly related to the acoustic environment—the soundscape— of the places in which we live, and that our memories and experiences of these places are mediated by the acoustic space (Böhme, 2000).

10.7 CONCLUSION

10.7.1 Urban Sound Planning: The Future

Soundscapes affect us, emotionally, physically, and psychologically, and each of us has our own experience of the sounds of the city, moulded by our cultural backgrounds, age, social relationships, and values. They help us to feel welcome or alienated in the places we live. As these wide-ranging examples illustrate, the soundscape approach, which has collaboration at its core, can effectively augment traditional noise mitigation strategies. The examples throughout this book, and more, illustrate an increasingly pro-active approach being taken by cities to manage local soundscapes as part of amenity, well-being, and environmental planning for residents and visitors. Even more, they provide opportunities for collaboration between scientists, urban planners, artists, architects, sound engineers, acousticians, and other practitioners. They demonstrate ways in which practitioners can apply sound-scape principles to achieve tangible improvements in people's experiences.

Most importantly, soundscape planning represents a much needed step change in the approach to managing the acoustic environment by putting people at the beginning of the process rather than the end. In this way, the users of a space become the de facto primary specifiers of any project. As Gernot Böhme (2000) noted, "City planning can no longer be content with noise control and abatement, but must pay attention to the character of the acoustic atmosphere of squares, pedestrian zones, of whole cities."

This outcomes-led approach is set to transform the role of acoustics professionals and noise control engineers from commodity suppliers to project partners. This evolution requires new skills, new insights, new standards, and new thinking from the industry as it currently exists and those pioneering the paths ahead. Indeed, the role of the hearing sense can no longer be relegated to mere preference, but must be understood as *primary* to how we experience the world around us.

In his book *The Soundscape: Our Sonic Environment and the Tuning of the World*, Canadian composer and polymath R. Murray Schafer (1994:3–4) states: "In various parts of the world important research is being undertaken ... dealing with aspects of the world soundscape, asking ... what is the relationship between [people] and the sound of [the] environment and what happens when those sounds change?" For the future, we will need more understanding, not just of how to control noise levels in decibel terms, but also of the qualitative aspects. This includes what the sounds of things mean to us in context. The paradigm shift that Schafer began in coining the term *soundscape* has far-reaching implications for a truly ecological approach by developing soundscapes that are supportive of human well-being.

The soundscape approach has much to offer city administrations as a complement to traditional strategies. The key is to take into account the opinions and experiences of the users of the space. The soundscape approach can help build social cohesion among citizens as they work together with experts to define the social use values of city spaces and agree together what sounds are appropriate and when, and how to make provision for one another's requirements; it can help to mitigate antisocial behaviour and promote safety, as well as quality of life through its ability to positively impact the psychological and physiological well-being of citizens; it can assist noise-sensitive and other vulnerable groups. It also respects cultural aspects and the beauty of natural or seminatural soundscapes.

With noise an ever-increasing problem in our towns and cities and traditional noise control measures proving increasingly difficult to apply and maintain, the emerging discipline of soundscape management offers policy makers a creative, inclusive, context-led approach to safeguarding sustainable communities for present and future generations. Central to the soundscape concept is its view of sound as a resource to be managed, protected, or enhanced, rather than noise as the waste by-product of a lack of soundscape management. As with the visual sphere, the adaptability of the soundscape offers limitless opportunities to designers when utilized at project planning phase.

And the time for this could not be better, with the recent development of the world's first international standard (ISO 12913-1:2014) to define, measure, and assess the acoustic environment using a soundscape approach. The wealth of examples representing the growing body of work on the topic, by experts from across the globe, bears testament that the development of such tools is enabling the prediction of annoyance, as well as a sense of well-being, based on the *sound quality* of environments—a feat not possible using only the decibel unit of measurement.

As noted in the introduction, this chapter has presented a range of examples representing recent work by participants rather than a comprehensive study. Taking the soundscape protocol (see Figure 10.1) into account with citizens at its heart, it is important to recognize the limitations of conventional questionnaire-type surveys when trying to capture subconscious effects, and emphasize that future work needs to continue to explore more behavioural observations in real-life experiments and more neurological modelling. But if you are reading this book, you have the power to make this happen. You have the power to design and build acoustically empathic places and spaces that will make the world a truly better place.

With this present state of the art, we have a unique opportunity to enter a new phase of management of acoustics in the built environment, to move beyond the blunt instruments of noise control and abatement to the wondrous exploration, endless creativity, and inclusiveness offered by soundscape design and management—making it possible to have buildings, communities, towns, and cities that literally resonate with well-being and amenity. Or, as Schafer (1994:245) admonishes, "Now it is our turn to anticipate what lies ahead of our ears and minds. You who would design the future world, listen forward … fifty, one hundred, a thousand years. What do you hear?"

REFERENCES

Anderson, S. (2013). Dublin City Acoustic Planning and Urban Sound Design Programme. In *Soundscape of European Cities and Landscapes*, ed. J. Kang, K. Chourmaouziadou, K. Sakantamis, B. Wang, and Y. Hao. Oxford: COST Office through Soundscape-COST, 344.

Anderson, S. (2014). MAP: Manual for acoustic planning and urban sound design. Dublin City Public Art Programme, Dublin City Council, Department of the Environment, Community and Local Government. http://map .minorarchitecture.org/ (accessed February 21, 2015).

Axelssön, Ö., ed. (2011). Designing soundscape for sustainable urban development. Stockholm, Stockholm. http://www.soundscape-conference.eu.

Bento Coelho, J.L., and Bañuelos, A. (2013). Urban redesign and sustainable mobility: An opportunity for soundscape in the urban future in Vitoria-Gasteiz. In *Soundscape of European Cities and Landscapes*, ed. J. Kang, K. Chourmaouziadou, K. Sakantamis, B. Wang, and Y. Hao. Oxford: COST Office through Soundscape-COST, 358.

Bohannon, R.W. (1997). Comfortable and maximum walking speed of adults aged 20–79 years: Reference values and determinants. *Age and Ageing*, 26(1), 15–19.

Böhme, G. (2000). Acoustic atmospheres: Contribution to the study of ecological aesthetics. *Soundscape: The Journal of Acoustic Ecology*, 1(1), 14–18.

British Standards Institute (BSI). (2014). BS 4142: 2014: Methods for rating and assessing industrial and commercial sound. London: BSI.

De Coensel, B., Bockstael, A., Dekoninck, L., Botteldooren, D., Schulte-Fortkamp, B., Kang, J., and Nilsson, M.E. (2010). The soundscape approach for early stage urban planning: A case study. Presented at 39th International Congress on Noise Control Engineering (Internoise 2010), Lisbon, Portugal.

De Coensel, B., and Botteldooren, D. (2010). Acoustic design for early stage urban planning. Presented at Proceedings of the International Conference on Designing Soundscapes for Sustainable Urban Development, Stockholm, Sweden.

Dökmeci, P., and Kang, J. (2013). Evaluating indoor soundscapes by questionnaires. In *Soundscape of European Cities and Landscapes*, ed. J. Kang, K. Chourmaouziadou, K. Sakantamis, B. Wang, and Y. Hao. Oxford: COST Office through Soundscape-COST, 345.

Drever, J.L. (2012). Sanitary ambiance: The noise effects of high speed hand dryers [project]. Goldsmiths Research Online. http://research.gold.ac.uk/7833/.

Easteal, M., Bannister, S., Kang, J., Aletta, F., Lavia, L., and Witchel, H. (2014). Urban sound planning in Brighton and Hove. Presented at Proceedings of Forum Acusticum 2014, Krakow, Poland.

García Pérez, I., and Aspuru Soloaga, I. (2013). Sound islands: Bilbao municipality policy to improve citizens quality of life through soundscape. In *Soundscape of European Cities and Landscapes*, ed. J. Kang, K. Chourmaouziadou, K. Sakantamis, B. Wang, and Y. Hao. Oxford: COST Office through Soundscape-COST, 350.

García Pérez, I., Aspuru Soloaga, I., Herranz-Pascual, K., and García-Borreguero, I. (2012). Validation of an indicator for the assessment of the environmental sound in urban places. Presented at Euronoise 2012, Prague.

Garrido, J.A., Siebein, G., and Ruiz, D.P. (2013). Soundwalks for the urban development of natural spaces in the Alachua County, Florida. In *Soundscape of European Cities and Landscapes,* ed. J. Kang, K. Chourmaouziadou, K. Sakantamis, B. Wang, and Y. Hao. Oxford: COST Office through Soundscape-COST, 354.

Herranz-Pascual, K., Iraurgi, I., García-Pérez, I., Aspuru, I., García-Borreguero, I., and Herrero-Fernández, D. (2011). Disruptive effect of urban environmental noise on the physiological recovery response after stress testing. Presented at International Congress on Noise as a Public Health Problem (ICBEN) 2, London.

International Organization for Standardization (ISO). (2014). ISO/FDIS 12913-1: 2014 Acoustics—Soundscape—Part 1: Definition and conceptual framework. Geneva: ISO.

Kang, J., Chourmaouziadou, K., Sakantamis, K., Wang, B., and Hao, Y., eds. (2013). *Soundscape of European Cities and Landscapes.* Oxford: COST Office through Soundscape-COST.

Kang, J., and Hao, Y. (2011). Waterscape and soundscape in Sheffield [Lecture]. Presented at Meeting of the COST Action TD0804 on Soundscape Examples in Community Context, Brighton, UK.

Lavia, L., Axelssön, Ö., and Dixon, M. (2012a). Sounding Brighton: Developing an applied soundscape strategy. In *AESOP 2012, Ankara,* ed. M. Balamir, M. Ersoy, and E. Babalik. Ankara: Sutcliffe Association of European Schools of Planning, paper 760.

Lavia, L., and Bennett, C., eds. (2011). *Soundscape: Where Life Sounds Good,* 2(40–43). http://noiseabatementsociety.com/soundscapeezine/past/ (accessed February 20, 2015).

Lavia, L., and Dixon, M. (2013). An exhibition exploring practical applications of soundscapes for the public realm. Presented at Euroregio Conference AIA-DAGA 2013, Merano, Italy.

Lavia, L., Easteal, M., Close, D., Witchel, H.J., Axelssön, O., Ware, M., and Dixon, M. (2012b). Sounding Brighton: Practical approaches towards better soundscapes. Presented at Internoise 2012, New York.

Lavia, L., Easteal, M., Close, D., Witchel, H.J., Axelssön, O., Ware, M., and Dixon, M. (2012c). West Street Story audio and video. For *Soundscapes of European Cities and Landscapes.* Oxford: COST. http://soundscape-cost.org/index.php?option=com_content&view=article&id=58:west-street-story&catid=35:soundscape-practices&Itemid=8 (accessed 21 February 2015).

Lopez-Mendez, A., Westling, C.E.I., Emonet, R., Easteal, M., Lavia, L., Witchel, H.J., and Odobez, J.-M. (2014). Automated bobbing and phase analysis to measure walking entrainment to music. Presented at IEEE International Conference on Image Processing, Paris.

Maffei, L. (2011). Soundscape approaches, including guidelines in soundscape heritages and tourism, drawing on experience in the historical town of Napoli, UNESCO patrimony [Lecture]. Presented at Brighton Soundscape Workshop from EU COST Action TD0804, Brighton and Hove City Council, and the UK Noise Abatement Society, Brighton and Hove, UK, April 6.

Maffei, L., and Di Gabriele, M. (2013). The cultural value of the soundscape of folk festivals. In *Soundscape of European Cities and Landscapes,* ed. J. Kang, K. Chourmaouziadou, K. Sakantamis, B. Wang, and Y. Hao. Oxford: COST Office through Soundscape-COST, 355.

Maffei, L., and Hiramatsu, K. (2006). A review of soundscape studies in Japan. *Acta Acustica united with Acustica*, 92(6), 857–866.

Ozcevik, A., and Can, Z.Y. (2013). An approach on the evaluation of urban acoustical comfort with the soundscape concept. In *Soundscape of European Cities and Landscapes*, ed. J. Kang, K. Chourmaouziadou, K. Sakantamis, B. Wang, and Y. Hao. Oxford: COST Office through Soundscape-COST, 340.

Schafer, R.M. (1977). *The Tuning of the World*. New York: Knopf.

Schafer, R.M. (1994). *The Soundscape: Our Sonic Environment and the Tuning of the World*. Rochester: Destiny Books. (First published 1977.) 245.

Schulte-Fortkamp, B. (2011). Soundscape approaches public space perception and enhancement drawing on experience in Berlin [Lecture]. Presented at Brighton Soundscape Workshop from COST Action TD0804, Brighton and Hove City Council, and the UK Noise Abatement Society, Brighton and Hove, April 6.

Schulte-Fortkamp, B. (2013). Soundscape Approaches public space perception and enhancement drawing on experience in Berlin. In *Soundscape of European Cities and Landscapes*, ed. J. Kang, K. Chourmaouziadou, K. Sakantamis, B. Wang, and Y. Hao. Oxford: COST Office through Soundscape-COST, 346.

Soundscape of European Cities and Landscapes Network. (2014). European Union COST Action TD 0804. Soundscapes of European Cities and Landscapes website. http://soundscape-cost.org/ (accessed February 20, 2015).

Ståhle, A. (2006). Sociotope mapping: Exploring public open space and its multiple use values in urban and landscape planning practice. *Nordic Journal of Architectural Research*, 19, 59–71.

Torija, A., and Ruiz, D.P. (2013). Soundscape-based criteria used in territorial planning: A case in the region of the Alpujarra, Granada, Spain. In *Soundscape of European Cities and Landscapes*, ed. J. Kang, K. Chourmaouziadou, K. Sakantamis, B. Wang, and Y. Hao. Oxford: COST Office through Soundscape-COST, 352.

UNESCO. (2003). Convention for the Safeguarding of the Intangible Cultural Heritage. Paris, October 17. http://www.unesco.org/culture/ich/RL/00869 (accessed February 21, 2015).

Vitoria-Gasteiz City Council. (2015). The greenbelt of Vitoria-Gasteiz. Vitoria-Gasteiz City Council. http://www.vitoria-gasteiz.org/ (accessed February 21, 2015).

Vitoria-Gasteiz City Council, Environmental Studies Centre. (2012). 1.3 green infrastructure function and benefits. In *The Interior Green Belt towards an Urban Green Infrastructure in Vitoria-Gasteiz*. Vitoria-Gasteiz, Spain: Vitoria-Gasteiz City Council.

Witchel, H. (2010). *You Are What You Hear: How Music and Territory Make Us Who We Are*. New York: Algora Publishing.

Witchel, H.J., Lavia, L., Easteal, M., Westling, C., Goodhand, D., Lopez-Mendez, A., and Odobez, J.-M. (2014). Music interventions in the West Street Tunnel in Brighton: A community safety and night-noise soundscape intervention pilot: Preliminary report. Brighton and Hove, UK: Brighton and Hove City Council.

Witchel, H.J., Lavia, L., Westling, C.E.I., Healy, A., Needham, R., and Chockalingam, N. (2013). Using body language indicators for assessing the effects of soundscape quality on individuals. Presented at AIA DAGA 2013, Merano, Italy.

Index